SpringerBriefs in Statistics

JSS Research Series in Statistics

The current research of statistics in Japan has expanded in several directions in line with recent trends in academic activities in the area of statistics and statistical sciences over the globe. The core of these research activities in statistics in Japan has been the Japan Statistical Society (JSS). This society, the oldest and largest academic organization for statistics in Japan, was founded in 1931 by a handful of pioneer statisticians and economists and now has a history of about 80 years. Many distinguished scholars have been members, including the influential statistician Hirotugu Akaike, who was a past president of JSS, and the notable mathematician Kiyosi Itô, who was an earlier member of the Institute of Statistical Mathematics (ISM), which has been a closely related organization since the establishment of ISM. The society has two academic journals: the Journal of the Japan Statistical Society (English Series) and the Journal of the Japan Statistical Society (Japanese Series). The membership of JSS consists of researchers, teachers, and professional statisticians in many different fields including mathematics, statistics, engineering, medical sciences, government statistics, economics, business, psychology, education, and many other natural, biological, and social sciences. The JSS Series of Statistics aims to publish recent results of current research activities in the areas of statistics and statistical sciences in Japan that otherwise would not be available in English; they are complementary to the two JSS academic journals, both English and Japanese. Because the scope of a research paper in academic journals inevitably has become narrowly focused and condensed in recent years, this series is intended to fill the gap between academic research activities and the form of a single academic paper. The series will be of great interest to a wide audience of researchers, teachers, professional statisticians, and graduate students in many countries who are interested in statistics and statistical sciences, in statistical theory, and in various areas of statistical applications.

Yuichi Goto · Hideaki Nagahata ·
Masanobu Taniguchi · Anna Clara Monti ·
Xiaofei Xu

ANOVA with Dependent Errors

Springer

Yuichi Goto
Faculty of Mathematics
Kyushu University
Fukuoka, Japan

Hideaki Nagahata
Risk Analysis Research Center
The Institute of Statistical Mathematics
Tachikawa, Tokyo, Japan

Masanobu Taniguchi
Waseda University
Shinjuku City, Tokyo, Japan

Anna Clara Monti
Department of Law, Economics
Management and Quantitative Methods
Università degli Studi del Sannio
Benevento, Italy

Xiaofei Xu
Department of Probability and Statistics
School of Mathematics and Statistics
Wuhan University
Wuhan, Hubei, China

ISSN 2191-544X ISSN 2191-5458 (electronic)
SpringerBriefs in Statistics
ISSN 2364-0057 ISSN 2364-0065 (electronic)
JSS Research Series in Statistics
ISBN 978-981-99-4171-1 ISBN 978-981-99-4172-8 (eBook)
https://doi.org/10.1007/978-981-99-4172-8

This Springer imprint is published by the registered company Springer Nature Singapore Pte Ltd.
The registered company address is: 152 Beach Road, #21-01/04 Gateway East, Singapore 189721, Singapore

To our families

Preface

The analysis of variance (ANOVA) is a statistical method for assessing the impact of multiple factors and their interactions when there are three or more factors. The method was first developed by R. A. Fisher in the 1910s and since then has been studied extensively by many authors.

In the case of i.i.d. data, most literature has focused on the setting where the number of groups (a) and the number of observations in each group (n) are small, referred to as fixed-a and -n asymptotics. ANOVA for time series data, commonly referred to as longitudinal or panel data analysis, has been extensively studied in econometrics. In this field, large-a and fixed-n asymptotics or large-a and -n asymptotics are commonly examined, with a primary focus on the regression coefficient. Consequently, results regarding the existence of fixed and random effects of factors have been hardly developed.

This monograph aims to present the recent developments related to one- and two-way models mainly for time series data under the framework of fixed-a and large-n asymptotics. Especially, we focus on (i) the testing problems for the existence of fixed and random effects of factors and interactions among factors under various settings, including uncorrelated and correlated groups, fixed and random effects, multi- and high-dimension, parametric and nonparametric spectral densities, and (ii) the local asymptotic normality (LAN) property for one-way models on i.i.d. data.

This book is suitable for statisticians and economists as well as psychologists and data analysts. Figure 1 illustrates the relationships between the chapters. In Chapter 1, a historical overview of ANOVA and the fundamentals of time series analysis are provided, along with motivation and concise summary of the content covered in the book. Chapter 2 examines a test for the presence of fixed effects in the one-way model with independent groups. Chapter 3 extends the analysis to high-dimensional settings. Chapters 4 and 5 address correlated groups in one-way and two-way models, respectively. Lastly, Chapter 6 explores the log-likelihood ratio process to construct optimal tests in the context of i.i.d. settings.

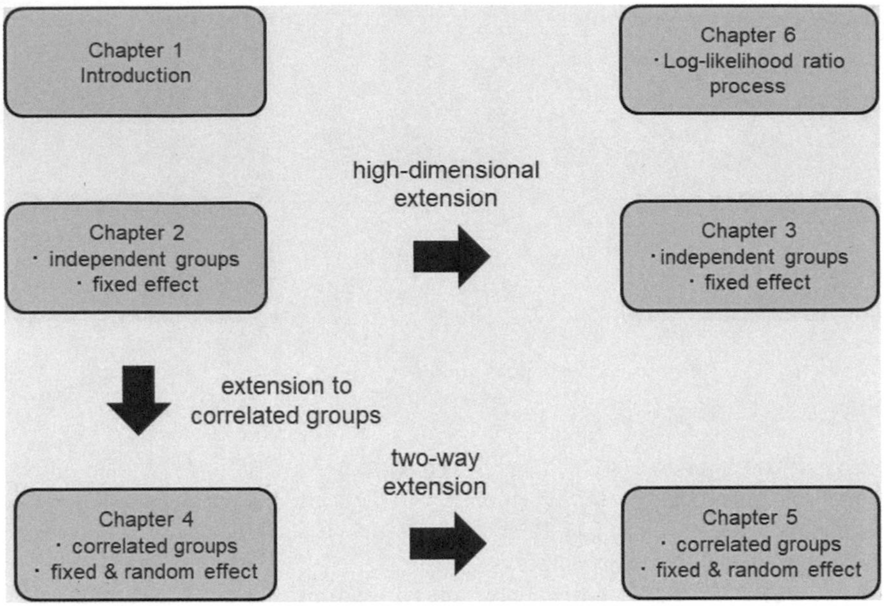

Fig. 1 Relationships between chapters

We are greatly indebted to Profs. M. Hallin, B. Shumway, D.S. Stoffer, C.R. Rao, P.M. Robinson, T. DiCiccio, S. Lee, C.W.S. Chen, Y. Chen, S. Yamashita, Y. Yajima, and Y. Matsuda for their valuable comments of fundamental impact on ANOVA and time series regression analysis. Thanks are extended to Drs. Y. Liu, F. Akashi, and Y. Xue for their collaboration and assistance with simulations.

The research was supported by JSPS Grant-in-Aid for Research Fellow under Grant Number JP201920060 (Y.G.); JSPS Grant-in-Aid for Research Activity Start-up under Grant Number JP21K20338 (Y.G.); JSPS Grant-in-Aid for Early-Career Scientists JP23K16851 (Y.G.); JSPS Grant-in-Aid for Early-Career Scientists JP20K13581 (H.N.); JSPS Grant-in-Aid for Challenging Exploratory Research under Grant Number JP26540015 (M.T.); JSPS Grant-in-Aid for Scientific Research (A) under Grant Number JP15H02061 (M.T.); JSPS Grant-in-Aid for Scientific Research (S) under Grant Number JP18H05290 (M.T.); the Research Institute for Science and Engineering (RISE) of Waseda University (M.T.); a start-up research grant of the Wuhan University under Grand Number 600460031 (X.X.).

Last but certainly not the least, we extend our sincerest appreciation to Mr. Yutaka Hirachi of Springer Japan and Mrs. Kavitha Palanisamy of Springer Nature for their assistance and patience.

Fukuoka, Japan	Yuichi Goto
Tokyo, Japan	Hideaki Nagahata
Tokyo, Japan	Masanobu Taniguchi
Benevento, Italy	Anna Clara Monti
Wuhan, China	Xiaofei Xu
February 2023	

Contents

Acronyms

$\mathbb{N}$	The set of natural numbers
$\mathbb{Z}$	The set of integers
$\mathbb{R}^d$	The set of d-tuples of real numbers
$A^\top$	The transpose of a matrix A
$\det(A)$	The determinant of a matrix A
$\mathbb{1}$	The indicator function
$\mathrm{tr}(A)$	The trace of a matrix A
A^*	The conjugate transpose of a matrix A
$\|A\|$	The Euclidean norm of a matrix A defined by $\sqrt{\mathrm{tr(A^*A)}}$
A^-	The Moore–Penrose inverse of a matrix A
$\mathbf{0}_p$	A p-dimensional vector whose entries are all zero
$\mathbf{1}_p$	A p-dimensional vector whose entries are all one
$\boldsymbol{O}_p$	A p-by-p zero matrix
$\boldsymbol{I}_p$	A p-by-p identity matrix
$\boldsymbol{J}_p$	A p-by-p matrix whose entries are all one
$X_n \to X$	In probability X_n converges in probability to X
$X_n \Rightarrow X$	X_n converges in distribution to X
$o_p(a_n)$	An order of the probability, that is, for a sequence of random variables $\{X_n\}$ and $\{a_n\}$,$0 < a_n \in \mathbb{R}$, $\left\{a_n^{-1}X_n\right\}$ converges in probability to zero
$O_p(a_n)$	An order of the probability, that is, for a sequence of random variables $\{X_n\}$ and $\{a_n\}$,$0 < a_n \in \mathbb{R}$, $\left\{a_n^{-1}X_n\right\}$ is bounded in probability
$O_p^U(\cdot)$	A p-by-p matrix whose elements are probability order $O_p(\cdot)$ with respect to all the elements uniformly
a and b	The number of groups
n_i	Sample size for an i-th group
n_{ij}	Sample size for an $(i,\ j)$-th group
n	Total sample size for all groups or cells
ρ_i and ρ_{ij}	A constant related to unbalanced data
φ	Significance level
$\chi_r^2[1-\varphi]$	The upper φ-percentiles of the chi-square distribution with r degrees of freedom

$\Psi_{r,\delta}$	The cumulative distribution function of the noncentral chi-square with r degrees of freedom and the noncentrality parameter δ
δ	The noncentrality parameter of the noncentral chi-square distribution
$\boldsymbol{Z}_{it}$	A t-th p-dimensional observation from an i-th group
$\boldsymbol{Z}_{ijt}$	A t-th p-dimensional observation from an (i, j)-th cell
$\boldsymbol{\mu}$	A p-dimensional general mean which is common to all groups or cells
$\boldsymbol{\alpha}_i$	A p-dimensional fixed or random effect for an i-th group or an i-th level of factor A
$\boldsymbol{\beta}_j$	A p-dimensional fixed or random effect for a j-th level of factor B
$\boldsymbol{\gamma}_{ij}$	A p-dimensional fixed or random interaction between the i-th level of factor A and the j-th level of factor B
${}^{\alpha}\Sigma$	The variance of random effects $(\alpha_1^\top, \ldots, \alpha_a^\top)^\top$
${}^{\beta}\Sigma$	The variance of random effects $(\beta_1^\top, \ldots \beta_b^\top)^\top$
${}^{\gamma}\Sigma$	The variance of random interactions $(\gamma_{11}^\top, \gamma_{21}^\top, \ldots, \gamma_{a1}^\top, \gamma_{12}^\top, \ldots, \gamma_{a2}^\top, \ldots, \gamma_{1b}^\top, \ldots, \gamma_{ab}^\top)^\top$
$\boldsymbol{e}_{it}$	A p-dimensional time series disturbance from the (i, j)-th cell at time t
$\boldsymbol{e}_{ijt}$	A spectral density matrix
$\boldsymbol{f}(\lambda)$	A spectral density matrix
$\hat{\boldsymbol{f}}_n(\lambda)$	Some consistent estimator of a spectral density matrix
$W(\cdot)$	A window function
$\omega(\cdot)$	A lag window function
M_n	The bandwidth parameter of the kernel method
$T_{\mathrm{LH},n}$	The classical Lawley–Hotelling test statistic, defined in (2.5), for independent observations for one-way model with the independent groups
$T_{\mathrm{LR},n}$	The classical likelihood ratio test statistic, defined in (2.6), for independent observations for one-way model with the independent groups
$T_{\mathrm{BNP},n}$	The classical Bartlett–Nanda–Pillai test statistic, defined in (2.7), for independent observations for one-way model with the independent groups
$T_{\mathrm{NT},n}$	The test statistic, defined in (2.16), for time series for one-way model with the independent groups
$T_{\mathrm{mLH},n,p}$	The modified Lawley–Hotelling test statistic to high-dimensional time series for one-way model with the independent groups defined in (3.10)
$T_{\mathrm{mLR},n,p}$	The modified likelihood ratio test statistic to high-dimensional time series for one-way model with the independent groups defined in (3.11)
$T_{\mathrm{mBNP},n,p}$	The modified Bartlett–Nanda–Pillai test statistic to high-dimensional time series for one-way model with the independent groups defined in (3.12)
$T_{\mathrm{iid},n}$	The classical F-statistic defined in (4.5)

$T_{\text{ts},n}$	The extended F-statistic to time series for one-way model with the independent groups defined in (4.6)
$T_{\text{GALT},n}$	The test statistic, defined in (4.7), for the existence of fixed and random effects for one-way model with correlated groups
$T_{\boldsymbol{\alpha},\text{GSXT},n}$	The test statistic, defined in (5.5), for the existence of random effects for two-way model with correlated groups
$T_{\boldsymbol{\gamma},\text{GSXT},n}$	The test statistic, defined in (5.8), for the existence of interactions for two-way model with correlated groups
$L_{\boldsymbol{Z}}$	The log-likelihood function of $\boldsymbol{Z}$ defined in (6.2)
$\Lambda(\boldsymbol{\theta}_0, \boldsymbol{\theta}_n)$	The log-likelihood ratio process for parameters $\boldsymbol{\theta}_0$ and $\boldsymbol{\theta}_n$ defined in (6.3)
$\boldsymbol{T}_{1,\text{GKKT},n}$	The quantity defined in (6.4) which appears in the asymptotic distribution of $\Lambda(\boldsymbol{\theta}_0, \boldsymbol{\theta}_n)$
$\boldsymbol{T}_{2,\text{GKKT},n}$	The quantity defined in (6.6) which appears in the asymptotic distribution of $\Lambda(\boldsymbol{\theta}_0, \boldsymbol{\theta}_n)$
$\phi_{\text{GKKT},n}$	The test function defined in (6.7)

Chapter 1
Introduction

This chapter describes elements of stationary processes. Concretely, we introduce the spectral distribution, orthogonal increment process, linear filter, and the response function. We explain the spectral representation of autocovariance functions and stationary process itself. Also, the relationship between linear filters and response functions is discussed. We mention a classical ANOVA approach in time series, and a brief outline of this book.

1.1 Foundations

Analysis of variance (ANOVA) has a long history. ANOVA deals with the problem of testing the null hypothesis that the means of different populations or the within-group means are all equal. The established theory covers testing and inference for independent observations (e.g., Rao, 1973; Anderson, 2003). However, recently, we observe dependent data in a variety of fields, e.g., finance, medical science, environmental science, engineering, and signal processing. For these dependent observations, we need the ANOVA approach, i.e., ANOVA with dependent errors. As an illustration, the plots G1–G3 in Fig. 1.1 show the financial returns for IBM, FORD, and Merck from February 12, 2021, to February 10, 2023, which are recognized as dependent data in view of substantial evidence. The data can be accessed from Yahoo Finance at https://finance.yahoo.com/.

For G_i, $i = 1, 2, 3$, we assume the following stochastic models:

$$G_i(t) = z_{it} = \mu_i + e_{it}, \quad i = 1, 2, 3, \tag{1.1}$$

where $\{e_{it}\}$'s are stationary processes, which will be detailed later. Consider the problem of testing the hypothesis:

$$H : \mu_1 = \mu_2 = \mu_3 \quad \text{versus} \quad K : H \text{ does not hold.}$$

Y. Goto et al., *ANOVA with Dependent Errors*, JSS Research Series in Statistics,
https://doi.org/10.1007/978-981-99-4172-8_1

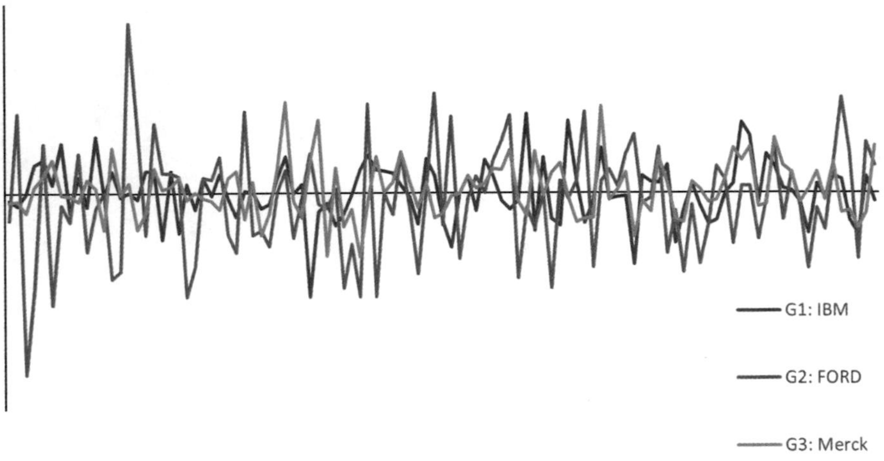

Fig. 1.1 Plots for the financial returns of IBM, FORD, and Merck

In Chapter 2, the test statistics will be introduced, and their statistical properties will be elucidated as well.

The model (1.1) is described by a stochastic process. In what follows, we provide its elements in vector form. Throughout the book, we use the following notations: $\mathbb{N}$ = the set of all positive integers, $\mathbb{Z}$ = the set of all integers, and $\mathbb{R}^m$ = the m-dimensional Euclidean space. Let $\{\boldsymbol{X}_t = (X_{1t}, \ldots, X_{mt})^\top : t \in \mathbb{Z}\}$ be a family of vector random variables, called an m-dimensional vector process, where $\top$ indicates transposition of the vector. The autocovariance function $\boldsymbol{\Gamma}(\cdot, \cdot)$ is defined by

$$\boldsymbol{\Gamma}(t, s) := \mathrm{Cov}(\boldsymbol{X}_t, \boldsymbol{X}_s) = \mathrm{E}\{(\boldsymbol{X}_t - \mathrm{E}\boldsymbol{X}_t)(\boldsymbol{X}_s - \mathrm{E}\boldsymbol{X}_s)^*\}, \quad t, s \in \mathbb{Z},$$

where $*$ denotes complex conjugate transpose. An m-dimensional vector process $\{\boldsymbol{X}_t : t \in \mathbb{Z}\}$ is said to be stationary if

(i) $\mathrm{E}(\boldsymbol{X}_t^* \boldsymbol{X}_t) < \infty$, for all $t \in \mathbb{Z}$;
(ii) $\mathrm{E}(\boldsymbol{X}_t) = \boldsymbol{c}$ for all $t \in \mathbb{Z}$, where $\boldsymbol{c}$ is a constant vector;
(iii) $\boldsymbol{\Gamma}(t, s) = \boldsymbol{\Gamma}(0, s - t)$ for all $s, t \in \mathbb{Z}$.

We state three fundamental theorems related to stationary processes (for the proofs, see, e.g., Hannan, 1970).

Theorem 1.1 *If $\boldsymbol{\Gamma}(\cdot)$ is the autocovariance function of an m-dimensional vector stationary process $\{\boldsymbol{X}_t\}$, then it has the representation*

$$\boldsymbol{\Gamma}(s) = \int_{-\pi}^{\pi} \mathrm{e}^{\mathrm{i}s\lambda} \mathrm{d}\boldsymbol{F}(\lambda), \quad s \in \mathbb{Z},$$

where $\boldsymbol{F}(\lambda)$ is a matrix-valued function whose increment $\boldsymbol{F}(\lambda_1) - \boldsymbol{F}(\lambda_2)$, $\lambda_1 \geq \lambda_2$, is non-negative definite. The function $\boldsymbol{F}(\lambda)$ is uniquely defined if in addition we require that *(i)* $\boldsymbol{F}(-\pi) = \boldsymbol{0}$*, and* $\boldsymbol{F}(\lambda)$ *is right continuous.*

The matrix $\boldsymbol{F}(\lambda)$ is called the spectral distribution matrix. If $\boldsymbol{F}(\lambda)$ is absolutely continuous with respect to the Lebesgue measure μ on $[-\pi, \pi]$ so that

$$\boldsymbol{F}(\lambda) = \int_{-\pi}^{\lambda} \boldsymbol{f}(\mu) \mathrm{d}\mu,$$

where $\boldsymbol{f}(\lambda)$ is a matrix with entries $f_{jk}(\lambda)$, then $\boldsymbol{f}(\lambda)$ is called the spectral density matrix. If

$$\sum_{h=-\infty}^{\infty} \|\boldsymbol{\Gamma}(h)\| < \infty,$$

where $\|\boldsymbol{\Gamma}(h)\| := \sqrt{\text{the greatest eigenvalue of } \boldsymbol{\Gamma}(h)^* \boldsymbol{\Gamma}(h)}$, the spectral density matrix is given by

$$\boldsymbol{f}(\lambda) = \frac{1}{2\pi} \sum_{h=-\infty}^{\infty} \boldsymbol{\Gamma}(h) \mathrm{e}^{-ih\lambda}.$$

For the spectral representation of $\{\boldsymbol{X}_t\}$, we need the concept of orthogonal increment process. We say that $\{\boldsymbol{Z}(\lambda) : -\pi \leq \lambda \leq \pi\}$ is an m-dimensional vector-valued orthogonal increment process if

(i) all the components of $\mathrm{E}\{\boldsymbol{Z}(\lambda)\boldsymbol{Z}(\lambda)^*\}$ are finite, $-\pi \leq \lambda \leq \pi$;
(ii) $\mathrm{E}\{\boldsymbol{Z}(\lambda)\} = \boldsymbol{0}, \ -\pi \leq \lambda \leq \pi$;
(iii) $\mathrm{E}[\{\boldsymbol{Z}(\lambda_4) - \boldsymbol{Z}(\lambda_3)\}\{\boldsymbol{Z}(\lambda_2) - \boldsymbol{Z}(\lambda_1)\}^*] = \boldsymbol{0}$ if $(\lambda_1, \lambda_2] \cap (\lambda_3, \lambda_4] = \varnothing$;
(iv) $\mathrm{E}[\{\boldsymbol{Z}(\lambda + \delta) - \boldsymbol{Z}(\lambda)\}\{\boldsymbol{Z}(\lambda + \delta) - \boldsymbol{Z}(\lambda)\}^*] \to 0$ as $\delta \to 0$.

Theorem 1.2 *If* $\{\boldsymbol{X}_t : t \in \mathbb{Z}\}$ *is an* m*-dimensional vector stationary process with* $\mathrm{E}(\boldsymbol{X}_t) = \boldsymbol{0}$ *and spectral distribution matrix* $\boldsymbol{F}(\lambda)$*, then there exists a right continuous orthogonal increment process* $\{\boldsymbol{Z}(\lambda) : -\pi \leq \lambda \leq \pi\}$ *such that*

(i) $\mathrm{E}[\{\boldsymbol{Z}(\lambda) - \boldsymbol{Z}(-\pi)\}\{\boldsymbol{Z}(\lambda) - \boldsymbol{Z}(-\pi)\}^*] = \boldsymbol{F}(\lambda), \ -\pi \leq \lambda \leq \pi$*;*
(ii)

$$\boldsymbol{X}_t = \int_{-\pi}^{\pi} \mathrm{e}^{-it\lambda} \mathrm{d}\boldsymbol{Z}(\lambda). \tag{1.2}$$

If a sequence of random vectors $\{\boldsymbol{a}_n\}$ satisfies $\|\boldsymbol{a}_n - \boldsymbol{a}\| \to 0$ as $n \to \infty$, for a constant vector $\boldsymbol{a}$, we say that $\{\boldsymbol{a}_n\}$ converges to the limit $\boldsymbol{a}$ in the mean. We denote this by $\mathrm{l.i.m.}_{n\to\infty} \boldsymbol{a}_n = \boldsymbol{a}$.

Theorem 1.3 *Suppose that* $\{\boldsymbol{X}_t : t \in \mathbb{Z}\}$ *is an* m*-dimensional vector stationary process with* $\mathrm{E}(\boldsymbol{X}_t) = \boldsymbol{0}$ *and spectral distribution matrix* $\boldsymbol{F}(\cdot)$ *and spectral representation (1.2), and that* $\{\boldsymbol{A}_j : j \in \mathbb{Z}\}$ *is a sequence of* $m \times m$ *matrices. Write*

$$\boldsymbol{Y}_t = \sum_{j=0}^{\infty} \boldsymbol{A}_j \boldsymbol{X}_{t-j}, \tag{1.3}$$

and

$$\boldsymbol{h}(\lambda) = \sum_{j=0}^{\infty} \boldsymbol{A}_j \mathrm{e}^{\mathrm{i}j\lambda}.$$

Then,

(i) the necessary and sufficient condition for (1.3) to exist as the l.i.m. of partial sums is

$$\mathrm{tr}\left\{\int_{-\pi}^{\pi} \boldsymbol{h}(\lambda)\mathrm{d}\boldsymbol{F}(\lambda)\boldsymbol{h}(\lambda)^*\right\} < \infty; \tag{1.4}$$

(ii) if (1.4) holds, the process $\{\boldsymbol{Y}_t : t \in \mathbb{Z}\}$ is stationary with the autocovariance function and spectral representation

$$\boldsymbol{\Gamma}_Y(s) = \int_{-\pi}^{\pi} \mathrm{e}^{\mathrm{i}s\lambda}\boldsymbol{h}(\lambda)\mathrm{d}\boldsymbol{F}(\lambda)\boldsymbol{h}(\lambda)^* \text{ and } \boldsymbol{Y}_t = \int_{-\pi}^{\pi} \mathrm{e}^{-\mathrm{i}t\lambda}\boldsymbol{h}(\lambda)\mathrm{d}\boldsymbol{Z}(\lambda),$$

respectively.

Let $\{\boldsymbol{A}_j\}$ be a sequence of $m \times m$-matrices satisfying

$$\sum_{j=0}^{\infty} \|\boldsymbol{A}_j\|^2 < \infty,$$

and let $\{\boldsymbol{U}_t : t \in \mathbb{Z}\}$ be a sequence of i.i.d. random vectors with mean zero and covariance matrix $\boldsymbol{V}$ (for short, $\{\boldsymbol{U}_t\} \sim i.i.d.(\boldsymbol{0}, \boldsymbol{V})$). Then, from Theorem 1.3, we can see that the generalized linear process

$$\boldsymbol{X}_t = \sum_{j=0}^{\infty} \boldsymbol{A}_j \boldsymbol{U}_{t-j} \tag{1.5}$$

is defined in the l.i.m. of partial sums, and that $\{\boldsymbol{X}_t\}$ has the spectral density matrix

$$\boldsymbol{f}(\lambda) = \frac{1}{2\pi}\left(\sum_{j=0}^{\infty} \boldsymbol{A}_j \mathrm{e}^{\mathrm{i}j\lambda}\right) \boldsymbol{V} \left(\sum_{j=0}^{\infty} \boldsymbol{A}_j \mathrm{e}^{\mathrm{i}j\lambda}\right)^*.$$

If $\{\boldsymbol{X}_t\}$ is generated by

$$\boldsymbol{X}_t + \boldsymbol{B}_1\boldsymbol{X}_{t-1} + \cdots + \boldsymbol{B}_p\boldsymbol{X}_{t-p} = \boldsymbol{U}_t + \boldsymbol{C}_1\boldsymbol{U}_{t-1} + \cdots + \boldsymbol{C}_q\boldsymbol{U}_{t-q},$$

where $\boldsymbol{B}_1, \ldots, \boldsymbol{B}_p, \boldsymbol{C}_1, \ldots, \boldsymbol{C}_q$ are $m \times m$ matrices, and $\{\boldsymbol{U}_t\} \sim i.i.d.(\boldsymbol{0}, \boldsymbol{V})$, it is called an m-dimensional vector autoregressive moving average of order (p, q) (for short VARMA(p, q)). Letting $\boldsymbol{B}(z) = I + \boldsymbol{B}_1 z + \cdots + \boldsymbol{B}_p z^p$, where I is the $m \times m$ identify matrix, we assume

$$\det \boldsymbol{B}(z) \neq 0 \text{ for all } z \in \mathbb{C} \text{ such that } |z| \leq 1.$$

Then, we can see that $\{\boldsymbol{X}_t\}$ is stationary and has the spectral density matrix

$$\boldsymbol{f}(\lambda) = \frac{1}{2\pi} \boldsymbol{B}\left(\mathrm{e}^{\mathrm{i}\lambda}\right)^{-1} \boldsymbol{C}\left(\mathrm{e}^{\mathrm{i}\lambda}\right) \boldsymbol{V} \boldsymbol{C}\left(\mathrm{e}^{\mathrm{i}\lambda}\right)^{*} \boldsymbol{B}\left(\mathrm{e}^{\mathrm{i}\lambda}\right)^{*-1}.$$

The VARMA(p, q) models are often used in many applications, and have the linear form of (1.5).

Consider the vector form of (1.1):

$$\boldsymbol{Z}_t = \boldsymbol{\mu} + \boldsymbol{e}_t, \tag{1.6}$$

where $\boldsymbol{Z}_t = (z_{1t}, z_{2t}, z_{3t})^\top$, $\boldsymbol{\mu} = (\mu_1, \mu_2, \mu_3)^\top$, and $\boldsymbol{e}_t = (e_{1t}, e_{2t}, e_{3t})^\top$. If we assume that $\{\boldsymbol{e}_t\}$ is stationary with spectral density matrix $\boldsymbol{f}(\lambda)$, we can apply the spectral analysis for (1.6).

Brillinger (1981) introduced the following model:

$$z_{jkt} = \mu + \alpha_t + \beta_{jt} + e_{jkt}, \tag{1.7}$$

where μ is a constant, $\{\alpha_t\}$ is a stationary process with $\mathrm{E}\{\alpha_t\} = 0$ and spectral density $f_{\alpha\alpha}(\lambda)$, $\{\beta_{jt}\}$, $j = 1, \ldots, J$ are stationary with $\mathrm{E}\{\beta_{jt}\} = 0$ and spectral density $f_{\beta\beta}(\lambda)$, and $\{e_{jkt}\}$, $j = 1, \ldots, J, k = 1, \ldots, K$, are stationary with $\mathrm{E}\{\varepsilon_{jkt}\} = 0$ and spectral density $f_{ee}(\lambda)$. For an observed stretch $\{X_{jkt}\}$, $t = 0, 1, \ldots, n-1$, let

$$d_{z_{jk}}^{(n)}(\lambda) := \sum_{t=0}^{n-1} z_{jkt} \mathrm{e}^{-\mathrm{i}\lambda t}, \quad d_{z_{j\cdot}}^{(n)}(\lambda) := \frac{1}{K} \sum_{k=1}^{K} d_{z_{jk}}^{(n)}(\lambda),$$

and

$$d_{z_{\cdot\cdot}}^{(n)}(\lambda) := \frac{1}{KJ} \sum_{k=1}^{K} \sum_{j=1}^{J} d_{z_{jk}}^{(n)}(\lambda).$$

Brillinger (1981) showed that

$$A_1(\lambda) := \frac{1}{J(K-1)} \sum_{j=1}^{J} \sum_{k=1}^{K} \frac{1}{2\pi n} \left| d_{z_{jk}}^{(n)}(\lambda) - d_{z_{j\cdot}}^{(n)}(\lambda) \right|^2,$$

$$A_2(\lambda) := J^{-1} K \sum_{j=1}^{J} \frac{1}{2\pi n} \left| d_{z_{j\cdot}}^{(n)}(\lambda) - d_{z_{\cdot\cdot}}^{(n)}(\lambda) \right|^2, \text{ and } A_3(\lambda) := \frac{1}{2\pi n} \left| d_{z_{\cdot\cdot\cdot}}^{(n)}(\lambda) \right|^2$$

are asymptotically, independently chi-square distributed. For each frequency λ, the statistics $A_1(\lambda)$, $A_2(\lambda)$, and $A_3(\lambda)$ can be used for the testing problems for the model (1.7).

Shumway and Stoffer (2006) introduced the following model:

$$z_{ijkt} = \mu_t + \alpha_{it} + \beta_{jt} + \gamma_{ijt} + e_{ijkt},$$

and, for each frequency λ, they developed a similar analysis.

In this book, we will develop systematic inference theories for time series ANOVA models by the use of the total information of the observed stretch instead of the information from each frequency.

1.2 Overview of Chapters

In the rest of this book, we will examine the following contents.

Chapter 2 addresses the problem of testing the null hypothesis H that the within-group means of time series ANOVA model are equal. For ANOVA with independent disturbances, the following tests, LR = likelihood ratio test, LH = Lawley–Hotelling test and BNP = Bartlett–Nanda–Pillai test, are illustrated.

For ANOVA with dependent disturbances, we will give a sufficient condition for LR, LH, and BNP to be asymptotically chi-square distributed under H. It is shown that generalized autoregressive conditional heteroscedasticity (GARCH) disturbance satisfies this sufficient condition.

Chapter 3 introduces p-dimensional ANOVA models

$$\boldsymbol{Z}_{it} = \boldsymbol{\mu} + \boldsymbol{\alpha}_i + \boldsymbol{e}_{it}, \quad i = 1, \dots, a,$$

and considers the problem of testing

$$H : \boldsymbol{\alpha}_1 = \cdots = \boldsymbol{\alpha}_a = \boldsymbol{0} \quad \text{v.s.} \quad K : \boldsymbol{\alpha}_i \neq 0 \text{ for some } i.$$

Let $T_i, i = 1, 2, 3$, be the standardized versions of LH, LR, and BNP, respectively. By assuming $p^{3/2}/\sqrt{n} \to 0$ as $n, p \to \infty$, we show that, under H, $T_i, i = 1, 2, 3$, follow asymptotically a standard normal distribution.

Chapter 4 introduces a random effect model

$$\boldsymbol{Z}_{it} = \boldsymbol{\mu} + \boldsymbol{\tau}_i + \boldsymbol{e}_{it}, \quad i = 1, \dots, a, \; t = 1, \dots, n_i,$$

where $\boldsymbol{Z}_{it}$ is a p-dimensional observation of the i-th group, $\boldsymbol{\mu}$ is the general mean, $\boldsymbol{\tau}_i$ is a p-dimensional normal random vector with mean $\boldsymbol{0}$ and variance matrix ${}^{\tau}\boldsymbol{\Sigma}$, and $(\boldsymbol{e}_{1t}^{\top}, \dots, \boldsymbol{e}_{at}^{\top})^{\top}$ is a sequence of stationary processes with spectral density matrix $\boldsymbol{f}(\lambda)$. We consider the problem of testing

$$H : {}^{\tau}\boldsymbol{\Sigma} = \boldsymbol{0} \text{ (no random effect) versus } K : {}^{\tau}\boldsymbol{\Sigma} \neq \boldsymbol{0} \text{ (random effect)},$$

and propose a new test $T_{\mathrm{GALT},n}$ with a quadratic form. Under H, it is shown that $T_{\mathrm{GALT},n}$ converges in a distribution to a chi-square distribution. Under K, we prove that $T_{\mathrm{GALT},n}$ is consistent. The results of Chapter 4 will be generalized to the case of two-way models in Chapter 5.

Chapter 6 explores the asymptotics of the likelihood ratio processes under non-standard settings. We consider a family of one-way random ANOVA models, and show that the model does not hold local asymptotic normality (LAN). Hence, we cannot apply the ordinary optimal theory based on LAN. However, we can show that the log-likelihood ratio test is asymptotically most powerful via Neyman–Pearson's lemma.

Chapter 7 deals with numerical studies for a variety of tests and models.

Chapter 8 provides the empirical data analysis. We assess the existence of area effects for the average wind speed data observed in seven cities located in Japan.

References

Anderson, T. W. (2003). *An introduction to multivariate statistical analysis* (3rd ed.). Wiley.
Brillinger, D. R. (1981). *Time series: Data analysis and theory*. San Francisco: Holden-Day.
Hannan, E. J. (1970). *Multiple time series*. Wiley.
Rao, C. R. (1973). *Linear statistical inference and its applications*. New York: Wiley.
Shumway, R. H., & Stoffer, D. S. (2006). *Time series analysis and its application with R examples* (2nd ed.). New York: Springer.

Chapter 2
One-Way Fixed Effect Model

While ANOVA for independent errors has been well-tuned, ANOVA for time-dependent errors is in its infancy. In this chapter, we extend the one-way fixed model with independent errors and groups to a one-way fixed model with time-dependent errors and independent groups. We illustrate the asymptotics of the classical test for time-dependent errors and propose the test for time-dependent errors introduced in Section 2.1. For the classical tests proposed for independent errors, we give sufficient conditions for their asymptotic chi-square distribution. For the case where sufficient conditions are violated, we propose to use a likelihood ratio test based on the Whittle likelihood. In what follows, we develop our discussion based on the results by Nagahata and Taniguchi (2017).

2.1 Tests for Time-Dependent Errors

In this section, we consider tests for the existence of fixed effects by introducing a one-way fixed effect model with time-dependent errors and independent groups.

Let $\boldsymbol{Z}_{i1}, \ldots, \boldsymbol{Z}_{in_i}$ be p-dimensional stretches, observed from the following one-way fixed model:

$$\boldsymbol{Z}_{it} = \boldsymbol{\mu} + \boldsymbol{\alpha}_i + \boldsymbol{e}_{it}, \; i = 1, \ldots, a, \; t = 1, \ldots, n_i, \tag{2.1}$$

where $\boldsymbol{\mu} = (\mu_1, \ldots, \mu_p)^\top$ is a grand mean, $\boldsymbol{\alpha}_i = (\alpha_{i1}, \ldots, \alpha_{ip})^\top$ is a fixed effect, and $\{\boldsymbol{e}_{it} = (e_{it1}, \ldots, e_{itp})^\top; \; i = 1, \ldots, a, \; t = 1, \ldots, n_i\}$ is the disturbance process. Here we assume $\sum_{i=1}^{a} \boldsymbol{\alpha}_i = \boldsymbol{0}_p$.

Suppose that

(i) the observed stretch $\{\boldsymbol{Z}_{it}; i = 1, \ldots, a, t = 1, \ldots, n_i\}$ is available.
(ii) the p-dimensional time series $\{\boldsymbol{e}_{it}\}$ is a centered stationary process, which has a p-by-p lag h autocovariance matrix $\boldsymbol{\Gamma}(h) = \{\Gamma_{j,k}(h)\,;\; j, k = 1, \ldots, p\}$, $h \in \mathbb{Z}$, and spectral density matrix $\boldsymbol{f}(\lambda) = (\boldsymbol{f}_i(\lambda))_{i=1,\ldots,a}$ for $\lambda \in [-\pi, \pi]$, where

Y. Goto et al., *ANOVA with Dependent Errors*, JSS Research Series in Statistics,
https://doi.org/10.1007/978-981-99-4172-8_2

$\boldsymbol{f}_i(\lambda)$ is a p-by-p spectrum of $\{\boldsymbol{e}_{it}\}$. Moreover, $\{\boldsymbol{e}_{i\cdot}\}$, $i = 1, \ldots, a$ are mutually independent.

(iii) $n_1 = \cdots = n_a$ with $n = \sum_{i=1}^{a} n_i$.

Remark 2.1 Condition (ii) is a standard assumption, called homoscedasticity (e.g., Section 8.9 in Anderson (2003)) and allows for the case of within-group correlation. In ANOVA and Design of Experiments, Condition (iii) is called balanced design.

Let $\{\boldsymbol{e}_{it}\}$ be generated from

$$\boldsymbol{e}_{it} = \sum_{j=0}^{\infty} \boldsymbol{A}_j \boldsymbol{\eta}_{i(t-j)}, \ \sum_{j=0}^{\infty} \|\boldsymbol{A}_j\|^2 < \infty, \ i = 1, \ldots, a, \tag{2.2}$$

where the $\boldsymbol{\eta}_{it}$'s are i.i.d. centered p-dimensional random vectors with variance $\boldsymbol{G}$, and the $\boldsymbol{A}_j$'s are $p \times p$ constant matrices. Then, $\{\boldsymbol{e}_{it}\}$ has the autocovariance matrices:

$$\boldsymbol{\Gamma}(l) = \sum_{j=0}^{\infty} \boldsymbol{A}_j \boldsymbol{G} \boldsymbol{A}_{j+l}^{\top},$$

and the spectral density matrix:

$$\boldsymbol{f}(\lambda) = \frac{1}{2\pi} \Big\{\sum_{j=0}^{\infty} \boldsymbol{A}_j e^{ij\lambda}\Big\} \boldsymbol{G} \Big\{\sum_{j=0}^{\infty} \boldsymbol{A}_j e^{ij\lambda}\Big\}^*.$$

We are interested in testing the hypothesis:

$$H_{21} : \boldsymbol{\alpha}_1 = \cdots = \boldsymbol{\alpha}_a \text{ versus } K_{21} : \ \boldsymbol{\alpha}_i \neq \boldsymbol{0} \text{ for some } i. \tag{2.3}$$

The null hypothesis H_{21} implies that all the effects are zero. Under the null hypothesis H_{21} defined in (2.3), we derive the asymptotic distribution of the test statistics.

2.1.1 Three Classical and Famous Test Statistics

In this subsection, we discuss the Lawley–Hotelling test, the likelihood ratio test, and the Bartlett–Nanda–Pillai test statistic proposed for independent observations.

To write the test statistics, we introduce

$$\hat{\boldsymbol{Z}}_{i\cdot} := \frac{1}{n_i} \sum_{t=1}^{n_i} \boldsymbol{Z}_{it}, \ \hat{\boldsymbol{Z}}_{\cdot\cdot} := \frac{1}{n} \sum_{i=1}^{a} \sum_{t=1}^{n_i} \boldsymbol{Z}_{it}, \tag{2.4}$$

$$\hat{\boldsymbol{\mathcal{S}}}_H := \sum_{i=1}^{a} n_i (\hat{\boldsymbol{Z}}_{i\cdot} - \hat{\boldsymbol{Z}}_{\cdot\cdot})(\hat{\boldsymbol{Z}}_{i\cdot} - \hat{\boldsymbol{Z}}_{\cdot\cdot})^{\top}, \text{ and } \hat{\boldsymbol{\mathcal{S}}}_E := \sum_{i=1}^{a} \sum_{t=1}^{n_i} (\boldsymbol{Z}_{it} - \hat{\boldsymbol{Z}}_{i\cdot})(\boldsymbol{Z}_{it} - \hat{\boldsymbol{Z}}_{i\cdot})^{\top},$$

where $n = n_1 + \cdots + n_a$. These are called within-group mean (for the i-th treatment group), grand mean, between-group sum of squares or sum of squares for treatment, and within-group sum of squares or sum of squares for error, respectively. If $\boldsymbol{e}_{it}$'s are mutually independent with respect to i, the following test statistics are proposed under normality:

$$T_{\mathrm{LH},n} := n\mathrm{tr}\{\hat{\boldsymbol{S}}_H \hat{\boldsymbol{S}}_E^{-1}\} \text{ (Lawley -- -Hotelling test)}, \tag{2.5}$$

$$T_{\mathrm{LR},n} := -n\log\{|\hat{\boldsymbol{S}}_E|/|\hat{\boldsymbol{S}}_E + \hat{\boldsymbol{S}}_H|\} \text{ (likelihood ratio test)}, \tag{2.6}$$

$$T_{\mathrm{BNP},n} := n\mathrm{tr}\hat{\boldsymbol{S}}_H(\hat{\boldsymbol{S}}_E + \hat{\boldsymbol{S}}_H)^{-1} \text{ (Bartlett–Nanda–Pillai test)}. \tag{2.7}$$

We consider the following assumption:

Assumption 2.1 $\det\{\boldsymbol{f}(0)\} > 0$.

Let $\hat{\boldsymbol{e}}_{i\cdot} := n_i^{-1}\sum_{t=1}^{n_i}\boldsymbol{e}_{it}$, and $\boldsymbol{e} := (\sqrt{n_1}\hat{\boldsymbol{e}}_{1\cdot}, \ldots, \sqrt{n_a}\hat{\boldsymbol{e}}_{a\cdot})$.

The following lemma is due to Hannan (1970) (p. 208, p. 221).

Lemma 2.1 *Under Assumption 2.1, as n_i, $i = 1, \ldots, a$, tends to ∞, if $\boldsymbol{e}_{it}$ is generated by the generalized linear process (2.2), then*

(i) $\hat{\boldsymbol{e}}_{i\cdot} \to \boldsymbol{0}_p$ in probability *for $i = 1, \ldots, a$;*
(ii) $\mathrm{vec}\{\boldsymbol{e}\}$ *converges in distribution to the ap-dimensional centered normal with variance*

$$2\pi\begin{pmatrix} \boldsymbol{f}(0) & \boldsymbol{O}_p & \cdots & \boldsymbol{O}_p \\ \boldsymbol{O}_p & \boldsymbol{f}(0) & \cdots & \boldsymbol{O}_p \\ \vdots & \vdots & & \vdots \\ \boldsymbol{O}_p & \boldsymbol{O}_p & \cdots & \boldsymbol{f}(0) \end{pmatrix}.$$

Theorem 2.1 *Assume that the processes $\{\boldsymbol{e}_{it}\}$ in (2.2) have the finite fourth-order cumulant, and that*

$$\boldsymbol{\Gamma}(j) = \boldsymbol{O}_p \ for\ all\ j \neq 0. \tag{2.8}$$

Then, under H_{21} and Assumption 2.1, all the tests $T_{\mathrm{LH},n}$, $T_{\mathrm{LR},n}$, and $T_{\mathrm{BNP},n}$ follow the asymptotic chi-square distribution with $(a-1)p$ degrees of freedom.

Proof (***Theorem*** 2.1) By the transformation $\boldsymbol{Z}_{it} \to \boldsymbol{\Gamma}(0)^{-1/2}\boldsymbol{Z}_{it}$, we observe that the three test statistics $T_{\mathrm{LH},n}$, $T_{\mathrm{LR},n}$, *and* $T_{\mathrm{BNP},n}$ are invariant. Thus, without loss of generality, we can assume $\boldsymbol{\Gamma}(0) = \boldsymbol{I}_p$. Under the setting of Section 2.1, we show that

$$\frac{1}{n}\hat{\boldsymbol{S}}_E = \boldsymbol{I}_p + O_p\left(\frac{1}{\sqrt{n_i}}\right). \tag{2.9}$$

Here, note that

$$T_{\mathrm{LH},n} = \mathrm{tr}\left\{\hat{\mathcal{S}}_H\left(\frac{1}{n}\hat{\mathcal{S}}_E\right)^{-1}\right\}, \tag{2.10}$$

$$T_{\mathrm{LR},n} = n \log \left\{ \left| \boldsymbol{I}_p + \hat{\mathcal{S}}_H \hat{\mathcal{S}}_E^{-1} \right| \right\}, \tag{2.11}$$

$$T_{\mathrm{BNP},n} = \mathrm{tr} \left\{ \hat{\mathcal{S}}_H \left(\frac{1}{n} \hat{\mathcal{S}}_E + \frac{1}{n} \hat{\mathcal{S}}_H \right)^{-1} \right\}. \tag{2.12}$$

By substituting (2.9) for (2.10), (2.11), and (2.12) and noting that $d \log |\boldsymbol{F}| = \mathrm{tr} \boldsymbol{F}^{-1} d\boldsymbol{F}$ (Magnus and Neudecker, 1999), we see that the stochastic expansion of the three statistics $T_{\mathrm{LH},n}, T_{\mathrm{LR},n}, T_{\mathrm{BNP},n} (= T)$ is given by

$$T = \mathrm{tr}\left(\boldsymbol{e} \boldsymbol{\Omega} \boldsymbol{e}^\top \right) + O_p \left(\frac{1}{\sqrt{n_i}} \right),$$

where $\boldsymbol{\Omega} = \boldsymbol{I}_a - \boldsymbol{\rho}\boldsymbol{\rho}^\top$, with $\boldsymbol{\rho} = (\sqrt{n_1/n}, \ldots, \sqrt{n_a/n})^\top$ (for i.i.d. case, e.g., Fujikoshi et al. 2011, pp. 164–165). Since $\boldsymbol{\Gamma}(j) = \boldsymbol{O}_p$, $(j \neq 0)$, Lemma 2.1 implies that $\mathrm{tr}\left(\boldsymbol{e} \boldsymbol{\Omega} \boldsymbol{e}^\top \right)$ has an asymptotic chi-square distribution with $(a-1)p$ degrees of freedom by Rao (2009). Thus, the results follow. □

Remark 2.2 The condition (2.8) implies that the $\{\boldsymbol{e}_{it}\}$'s are uncorrelated processes. Theorem 2.1 shows that the three test statistics proposed for independent observations apply also to dependent observations satisfying (2.8). Note that the condition (2.8) is not very stringent since the following practical nonlinear time series model satisfies (2.8). Bollerslev et al. (1988) introduced the vector generalized autoregressive conditional heteroscedasticity (GARCH(u, v)) model:

$$\boldsymbol{e}_t = \boldsymbol{H}_t^{1/2} \boldsymbol{\eta}_t \text{ and } \mathrm{vech}(\boldsymbol{H}_t) = \boldsymbol{w} + \sum_{i=1}^{u} \boldsymbol{B}_i \mathrm{vech}\left(\boldsymbol{H}_{t-i} \right) + \sum_{j=1}^{v} \boldsymbol{C}_j \mathrm{vech}\left(\boldsymbol{e}_{t-j} \boldsymbol{e}_{t-j}^\top \right),$$

where $\boldsymbol{e}_t = (e_{t1}, \ldots, e_{tp})^\top$, $\{\boldsymbol{\eta}_t; t = 1, 2, \ldots\}$ follows the i.i.d. centered p-dimensional vector with variance $\boldsymbol{I}_p$, vech($\cdot$) denotes the column stacking operator of the lower portion of a symmetric matrix, $\boldsymbol{w}$ is $p(p+1)/2$ constant vector, and $\boldsymbol{B}_i$s and $\boldsymbol{C}_j$s are $(p(p+1)/2) \times (p(p+1)/2)$ constant matrices. Let $\mathcal{F}_{t-1}$ be the σ-algebra generated by $\{\boldsymbol{e}_{t-1}, \boldsymbol{e}_{t-2}, \ldots\}$. We assume $\boldsymbol{H}_t$ is measurable with respect to $\mathcal{F}_{t-1}$ and $\boldsymbol{\eta}_t \perp \mathcal{F}_{t-1}$. Since this model is critical and widespread in nonlinear time series analysis of financial data, the three tests based on $T_{\mathrm{LH},n}$, $T_{\mathrm{LR},n}$, *and* $T_{\mathrm{BNP},n}$ can be applied to financial data.

2.1.2 Likelihood Ratio Test Based on Whittle Likelihood

In Theorem 2.1, we saw that the classical test statistics $T_{\mathrm{LH},n}$, $T_{\mathrm{LR},n}$, *and* $T_{\mathrm{BNP},n}$ are asymptotically chi-square distributed when (2.8) holds. However, the tests $T_{\mathrm{LH},n}$, $T_{\mathrm{LR},n}$, *and* $T_{\mathrm{BNP},n}$ are not available if one wants to test the hypothesis defined in (2.3) for general disturbances which do not satisfy (2.8). For this reason, we propose a new test based on the Whittle likelihood.

Whittle's approximation to the Gaussian likelihood function is given by

$$l(\boldsymbol{\mu}, \alpha) := -\frac{1}{2}\sum_{i=1}^{a}\sum_{s=0}^{n_i-1} \operatorname{tr}\left\{\boldsymbol{I}_i(\lambda_s)\boldsymbol{f}(\lambda_s)^{-1}\right\},$$

where $\lambda_s = 2\pi s/n_i$ and

$$\boldsymbol{I}_i(\lambda) := \frac{1}{2\pi n_i}\left\{\sum_{t=1}^{n_i}(\boldsymbol{Z}_{it} - \boldsymbol{\mu} - \boldsymbol{\alpha}_i)\mathrm{e}^{\mathrm{i}\lambda t}\right\}\left\{\sum_{u=1}^{n_i}(\boldsymbol{Z}_{iu} - \boldsymbol{\mu} - \boldsymbol{\alpha}_i)\mathrm{e}^{\mathrm{i}\lambda u}\right\}^{*}.$$

Under the null hypothesis H_{21} defined in (2.3), we know from $\frac{\partial l(\boldsymbol{\mu},\boldsymbol{0}_p)}{\partial \boldsymbol{\mu}} = \boldsymbol{0}_p$, $\frac{\partial l(\boldsymbol{\mu},\boldsymbol{\alpha})}{\partial \boldsymbol{\mu}} = \boldsymbol{0}_p$, $\frac{\partial l(\boldsymbol{\mu},\boldsymbol{\alpha})}{\partial \boldsymbol{\alpha}_i} = \boldsymbol{0}_p$ that the solutions are

$$\boldsymbol{\mu} = \hat{\boldsymbol{\mu}}_{\cdot\cdot} := \frac{1}{n}\sum_{i=1}^{a}\sum_{t=1}^{n_i}\boldsymbol{Z}_{it} \text{ and } \boldsymbol{\alpha} = \hat{\boldsymbol{\alpha}}_{i\cdot} := \frac{1}{n_i}\sum_{t=1}^{n_i}(\boldsymbol{Z}_{it} - \hat{\boldsymbol{\mu}}_{\cdot\cdot}).$$

Hence, we introduce the following test statistic:

$$T_{\mathrm{WLR}} := 2\left\{l(\hat{\boldsymbol{\mu}}_{\cdot\cdot}, \hat{\boldsymbol{\alpha}}) - l(\hat{\boldsymbol{\mu}}_{\cdot\cdot}, \boldsymbol{0}_p)\right\},$$

or equivalently

$$T_{\mathrm{WLR}} = \sum_{i=1}^{a}\sqrt{n_i}\hat{\boldsymbol{\alpha}}_{i\cdot}^{\top}\left\{2\pi \boldsymbol{f}(0)\right\}^{-1}\sqrt{n_i}\hat{\boldsymbol{\alpha}}_{i\cdot}. \tag{2.13}$$

The following result holds.

Lemma 2.2 *Suppose Assumption 2.1 holds. If $\boldsymbol{e}_{it}$ is generated by the generalized linear process (2.2), then under H_{21}, the test statistic T_{WLR} has an asymptotic chi-square distribution with $(a-1)p$ degrees of freedom.*

Proof (***Lemma*** 2.2) Under H_{21}, we obtain

$$\begin{aligned}\frac{\partial l(\boldsymbol{\mu}, \boldsymbol{0}_p)}{\partial \boldsymbol{\mu}} = -\frac{1}{2}\sum_{i=1}^{a}\Bigg[\sum_{s=0}^{n_i-1}\boldsymbol{f}(\lambda_s)^{-1}\Bigg\{&\frac{1}{2\pi n_i}\sum_{t=1}^{n_i}(-\mathrm{e}^{\mathrm{i}\lambda_s t})\sum_{u=1}^{n_i}(\boldsymbol{Z}_{iu} - \boldsymbol{\mu})\mathrm{e}^{-\mathrm{i}\lambda_s u}\\ &+ \frac{1}{2\pi n_i}\sum_{t=1}^{n_i}(\boldsymbol{Z}_{it} - \boldsymbol{\mu})\mathrm{e}^{\mathrm{i}\lambda_s t}\sum_{u=1}^{n_i}(-\mathrm{e}^{-\mathrm{i}\lambda_s u})\Bigg\}\Bigg].\end{aligned}$$

Noting that

$$\frac{1}{n_i}\sum_{t=1}^{n_i}\mathrm{e}^{i\lambda_s t} = \begin{cases}1 \ (s=0)\\ 0 \ (s \neq 0),\end{cases} \tag{2.14}$$

we can see that $\frac{\partial l(\boldsymbol{\mu},\mathbf{0}_p)}{\partial \boldsymbol{\mu}} = \mathbf{0}_p$ leads to the solution

$$\boldsymbol{\mu} = \frac{1}{n}\sum_{i=1}^{a}\sum_{t=1}^{n_i}\mathbf{Z}_{it} \, .$$

Next,

$$\frac{\partial l(\boldsymbol{\mu},\boldsymbol{\alpha})}{\partial \boldsymbol{\alpha}_i} = -\frac{1}{2}\sum_{s=0}^{n_i-1} \boldsymbol{f}(\lambda_s)^{-1}\left\{\frac{1}{2\pi n_i}\sum_{t=1}^{n_i}(-\mathrm{e}^{\mathrm{i}\lambda_s t})\sum_{u=1}^{n_i}(\mathbf{Z}_{iu} - \boldsymbol{\mu} - \boldsymbol{\alpha}_i)\mathrm{e}^{-\mathrm{i}\lambda_s u}\right.$$
$$\left. + \frac{1}{2\pi n_i}\sum_{t=1}^{n_i}(\mathbf{Z}_{it} - \boldsymbol{\mu} - \boldsymbol{\alpha}_i)\mathrm{e}^{\mathrm{i}\lambda_s t}\sum_{u=1}^{n_i}(-\mathrm{e}^{-\mathrm{i}\lambda_s u})\right\} = \mathbf{0}_p$$

leads to

$$\boldsymbol{\alpha}_i = \frac{1}{n_i}\sum_{t=1}^{n_i}(\mathbf{Z}_{it} - \boldsymbol{\mu}).$$

Similarly, from $\frac{\partial l(\boldsymbol{\mu},\boldsymbol{\alpha})}{\partial \boldsymbol{\mu}} = \mathbf{0}_p$, we obtain

$$\boldsymbol{\mu} = \frac{1}{n}\sum_{i=1}^{a}\sum_{t=1}^{n_i}(\mathbf{Z}_{it} - \boldsymbol{\alpha}_i).$$

As a solution, we may take

$$\boldsymbol{\alpha} = \hat{\boldsymbol{\alpha}}_{i\cdot} := \frac{1}{n_i}\sum_{t=1}^{n_i}(\mathbf{Z}_{it} - \hat{\boldsymbol{\mu}}_{\cdot\cdot}) \text{ and } \boldsymbol{\mu} = \hat{\boldsymbol{\mu}}_{\cdot\cdot} := \frac{1}{n}\sum_{i=1}^{a}\sum_{t=1}^{n_i}\mathbf{Z}_{it}.$$

From the above, it follows that

$$\begin{aligned}
T_{\mathrm{WLR}} &= 2\left\{l(\hat{\boldsymbol{\mu}}_{\cdot\cdot}, \hat{\boldsymbol{\alpha}}) - l(\hat{\boldsymbol{\mu}}_{\cdot\cdot}, \mathbf{0}_p)\right\} \\
&= \sum_{i=1}^{a}\sum_{s=0}^{n_i-1}\mathrm{tr}\left\{\frac{1}{2\pi n_i}\sum_{t=1}^{n_i}(\mathbf{Z}_{it} - \hat{\boldsymbol{\mu}}_{\cdot\cdot})\mathrm{e}^{\mathrm{i}\lambda_s t}\sum_{u=1}^{n_i}(\mathbf{Z}_{iu} - \hat{\boldsymbol{\mu}}_{\cdot\cdot})^\top \mathrm{e}^{-\mathrm{i}\lambda_s u}\boldsymbol{f}(\lambda_s)^{-1}\right. \\
&\qquad \left. - \frac{1}{2\pi n_i}\sum_{t=1}^{n_i}(\mathbf{Z}_{it} - \hat{\boldsymbol{\mu}}_{\cdot\cdot} - \hat{\boldsymbol{\alpha}}_{i\cdot})\mathrm{e}^{\mathrm{i}\lambda_s t}\sum_{u=1}^{n_i}(\mathbf{Z}_{iu} - \hat{\boldsymbol{\mu}}_{\cdot\cdot} - \hat{\boldsymbol{\alpha}}_{i\cdot})^\top \mathrm{e}^{-\mathrm{i}\lambda_s u}\boldsymbol{f}(\lambda_s)^{-1}\right\} \\
&= \sum_{i=1}^{a}\sum_{s=0}^{n_i-1}\mathrm{tr}\left[-\frac{1}{2\pi n_i}\sum_{t=1}^{n_i}\hat{\boldsymbol{\alpha}}_{i\cdot}\mathrm{e}^{\mathrm{i}\lambda_s t}\sum_{u=1}^{n_i}\hat{\boldsymbol{\alpha}}_{i\cdot}^\top \mathrm{e}^{\mathrm{i}\lambda_s u}\boldsymbol{f}(\lambda_s)^{-1}\right. \\
&\qquad +\frac{1}{2\pi n_i}\sum_{t=1}^{n_i}\hat{\boldsymbol{\alpha}}_{i\cdot}\mathrm{e}^{\mathrm{i}\lambda_s t}\sum_{u=1}^{n_i}(\mathbf{Z}_{iu} - \hat{\boldsymbol{\mu}}_{\cdot\cdot})^\top \mathrm{e}^{-\mathrm{i}\lambda_s u}\boldsymbol{f}(\lambda_s){-}1
\end{aligned}$$

$$+\frac{1}{2\pi n_i}\sum_{t=1}^{n_i}(\mathbf{Z}_{it}-\hat{\boldsymbol{\mu}}_{..})e^{i\lambda_s t}\sum_{u=1}^{n_i}\hat{\boldsymbol{\alpha}}_i^\top e^{-i\lambda_s u}\boldsymbol{f}(\lambda_s)^{-1}\Bigg].$$

Recalling (2.14), we obtain

$$\begin{aligned}
&T_{\mathrm{WLR}}\\
&=\sum_{i=1}^{a}\mathrm{tr}\Bigg[-\frac{1}{2\pi}\hat{\boldsymbol{\alpha}}_{i\cdot}\sum_{u=1}^{n_i}\hat{\boldsymbol{\alpha}}_{i\cdot}^\top\boldsymbol{f}(0)^{-1}+\hat{\boldsymbol{\alpha}}_{i\cdot}\frac{1}{2\pi}\sum_{u=1}^{n_i}(\mathbf{Z}_{iu}-\hat{\boldsymbol{\mu}}_{..})^\top\boldsymbol{f}(0)^{-1}\\
&\qquad+\frac{1}{2\pi}\sum_{t=1}^{n_i}(\mathbf{Z}_{it}-\hat{\boldsymbol{\mu}}_{..})\hat{\boldsymbol{\alpha}}_{i\cdot}^\top\boldsymbol{f}(0)^{-1}\Bigg]\\
&=\sum_{i=1}^{a}\mathrm{tr}\left[-\hat{\boldsymbol{\alpha}}_{i\cdot}n_i\hat{\boldsymbol{\alpha}}_{i\cdot}^\top\{2\pi\boldsymbol{f}(0)\}^{-1}+2\hat{\boldsymbol{\alpha}}_{i\cdot}n_i\hat{\boldsymbol{\alpha}}_{i\cdot}^\top\{2\pi\boldsymbol{f}(0)\}^{-1}\right]\\
&=\sum_{i=1}^{a}\sqrt{n_i}\hat{\boldsymbol{\alpha}}_{i\cdot}^\top\{2\pi\boldsymbol{f}(0)\}^{-1}\sqrt{n_i}\hat{\boldsymbol{\alpha}}_{i\cdot}.
\end{aligned}\tag{2.15}$$

Note that

$$\hat{\boldsymbol{\alpha}}_{i\cdot}=\bar{\boldsymbol{\alpha}}_{i\cdot}-\frac{1}{n}\sum_{i=1}^{a}n_i\bar{\boldsymbol{\alpha}}_{i\cdot},\quad \text{where } \bar{\boldsymbol{\alpha}}_{i\cdot}=\frac{1}{n_i}\sum_{t=1}^{n_i}(\mathbf{Z}_{it}-\boldsymbol{\mu}).$$

Expression (2.15) is a mean-corrected quadratic form. Since $\sqrt{n_i}\bar{\boldsymbol{\alpha}}_{i\cdot}$ converges in distribution to the p-dimensional centered normal distribution with variance $2\pi\boldsymbol{f}(0)$ (see Hannan, 1970, p. 208), we can see that T_{WLR} converges in distribution to the chi-square distribution with $(a-1)p$ degrees of freedom.

Remark 2.3 As Lemma 2.2, the new test statistic based on the Whittle likelihood is asymptotically chi-square distributed even if the condition (2.8) does not hold.

Furthermore, to propose a practical version of T_{WLR} we consider

$$\hat{\boldsymbol{f}}_i(\lambda):=\frac{1}{2\pi}\sum_{h=-(n_i-1)}^{n_i-1}\omega\left(\frac{h}{M_{n_i}}\right)e^{-ih\lambda}\left(1-\frac{|h|}{n_i}\right)\hat{\boldsymbol{\Gamma}}_i(h),$$

$$\hat{\boldsymbol{\Gamma}}_i(h):=\frac{1}{n_i}\sum_{t=1}^{n_i-h}(\mathbf{Z}_{i(t+h)}-\hat{\mathbf{Z}}_{i\cdot})(\mathbf{Z}_{it}-\hat{\mathbf{Z}}_{i\cdot})^\top,$$

where M_{n_i} is a positive sequence of integers, $\hat{\mathbf{Z}}_{i\cdot}$ in $\hat{\boldsymbol{\Gamma}}_i(h)$ is defined in (2.4), $\omega(x):=\int_{-\infty}^{\infty}W(t)e^{ixt}dt$, and the function $W(\cdot)$ satisfies the following conditions: $W(\cdot)$ is a real, bounded, non-negative, even function such that $\int_{-\infty}^{\infty}W(t)dt=1$ and $\int_{-\infty}^{\infty}W^2(t)dt<\infty$ with a bounded derivative.

The following assumption is made.

Assumption 2.2 For the same integer ν,

(i) for some $\nu \geq 1$,

$$\lim_{x \to 0} \frac{1 - \omega(x)}{|x|^{\nu}} < \infty,$$

(ii) $M_{n_i} \to \infty$, $\{M_{n_i}\}^{\nu}/n_i \to 0$,

(iii)

$$\sum_{s_2=-\infty}^{\infty} \sum_{s_3=-\infty}^{\infty} \sum_{s_4=-\infty}^{\infty} |\kappa^{i}_{r_1,r_2,r_3,r_4}(s_2, s_3, s_4)| < \infty, \quad \sum_{h=-\infty}^{\infty} |h|^{\nu} ||\boldsymbol{\Gamma}(h)|| < \infty, \ \nu \geq 0$$

where $\kappa^{i}_{r_1,r_2,r_3,r_4}(s_2, s_3, s_4) := \mathrm{Cum}\{e_{i0r_1}, e_{is_2r_2}, e_{is_3r_3}, e_{is_4r_4}\}$ for $i = 1, \ldots, a$ and $r_1, r_2, r_3, r_4 = 1, \ldots, p$.

The following lemma is due to Hannan (1970) (p. 280, p. 283, and p. 331).

Lemma 2.3 *Under Assumption 2.2, as* n_i, $i = 1, \ldots, a$, *tend to* ∞, $\hat{\boldsymbol{f}}_i(\lambda) \to \boldsymbol{f}(\lambda)$ in probability *for* $i = 1, \ldots, a$.

Under Assumption 2.2, we can replace $\boldsymbol{f}(0)$ in (2.13) by $\hat{\boldsymbol{f}}_i(0)$:

$$T_{\mathrm{NT},n} = \sum_{i=1}^{a} \sqrt{n_i} \hat{\boldsymbol{\alpha}}_{i\cdot}^{\top} \left\{ 2\pi \hat{\boldsymbol{f}}_i(0) \right\}^{-1} \sqrt{n_i} \hat{\boldsymbol{\alpha}}_{i\cdot}. \tag{2.16}$$

Therefore, the following result immediately follows from Slutsky's theorem.

Theorem 2.2 *Under* H_{21} *and Assumptions 2.1 and 2.2, if* $\mathbf{e}_{it}$ *is generated by the generalized linear process (2.2), then the test statistic* $T_{\mathrm{NT},n}$ *has an asymptotic chi-square distribution with* $(a-1)p$ *degrees of freedom.*

Remark 2.4 The test statistic $T_{\mathrm{NT},n}$ in Theorem 2.2 is practically useful because it can be calculated directly from the observed values.

References

Anderson, T. W. (2003). *An introduction to multivariate statistical analysis* (3rd ed.). Wiley.

Bollerslev, T., Engle, R. F., & Wooldridge, J. M. (1988). A capital asset pricing model with time-varying covariances. *The Journal of Political Economy*, 116–131.

Fujikoshi, Y., Ulyanov, V. V., & Shimizu, R. (2011). *Multivariate statistics: High-dimensional and large-sample approximations 760*. Wiley.

Hannan, E. J. (1970). *Multiple time series*. Wiley.

Magnus, J. R., & Neudecker, H. (1999). *Matrix differential calculus with applications in statistics and econometrics*. New York: Wiley.
Nagahata, H., & Taniguchi, M. (2017). Analysis of variance for multivariate time series. *Metron, 76,* 69–82.
Rao, C. R. (2009). *Linear statistical inference and its applications 22*. Wiley.

Chapter 3
One-Way Fixed Effect Model for High-Dimensional Time Series

Although finite dimensionality was assumed in Chapter 2, ANOVA for high-dimensional time-dependent errors has not been fully developed. In this chapter, we extend the one-way fixed model with time-dependent errors and independent groups to a one-way fixed model with high-dimensional time-dependent errors and independent groups. In Section 3.1, we develop the asymptotics of the basic statistics and modified classical tests for high-dimensional time-dependent errors. A sufficient condition for the modified classical tests to be asymptotically normal is also presented. This chapter is based mostly on Nagahata and Taniguchi (2018).

3.1 Tests for High-Dimensional Time-Dependent Errors

Throughout, we consider the one-way fixed effect model under which a a-tuple of p-dimensional time series $\boldsymbol{Z}_{i1}, \dots, \boldsymbol{Z}_{in_i}$, $i = 1, \dots, a$ satisfies

$$\boldsymbol{Z}_{it} = \boldsymbol{\mu} + \boldsymbol{\alpha}_i + \boldsymbol{e}_{it}, \quad i = 1, \dots, a, \ t = 1, \dots, n_i, \tag{3.1}$$

where the disturbances $\boldsymbol{e}_{it} = (e_{it1}, \dots, e_{itp})^\top$. Here, $\boldsymbol{\mu}$ is the global mean of the model (3.1), and $\boldsymbol{\alpha}_i$ denotes the fixed effects of the i-th treatment, which measures the deviation from the grand mean $\boldsymbol{\mu}$ satisfying $\sum_{i=1}^a \boldsymbol{\alpha}_i = \boldsymbol{0}_p$.

We suppose that

(i) the observed stretch $\{\boldsymbol{Z}_{it}; i = 1, \dots, a, t = 1, \dots, n_i\}$ is available,
(ii) the p-dimensional time series $\{\boldsymbol{e}_{it}\}$ is a centered stationary process, which has a p-by-p lag h autocovariance matrix $\boldsymbol{\Gamma}(h) = \left(\Gamma_{jk}(h)\right)_{1 \le j,k \le p}$, $h \in \mathbb{Z}$ and spectral density matrix $\boldsymbol{f}(\lambda) = (\boldsymbol{f}_i(\lambda))_{i=1,\dots,a}$ for $\lambda \in [-\pi, \pi]$, where $\boldsymbol{f}_i(\lambda)$ is a p-by-p spectrum of $\{\boldsymbol{e}_{it}\}$, furthermore $\{\boldsymbol{e}_{i\cdot}\}$, $i = 1, \dots, a$ are mutually independent, and
(iii) $n_1 = \cdots = n_a$ with $n = \sum_{i=1}^a n_i$.

Y. Goto et al., *ANOVA with Dependent Errors*, JSS Research Series in Statistics,
https://doi.org/10.1007/978-981-99-4172-8_3

Remark 3.1 The above conditions are the same as those considered at the beginning of Chapter 2.

Since the sum of the treatment effects is zero, we consider the hypotheses:

$$H_{31}: \boldsymbol{\alpha}_1 = \cdots = \boldsymbol{\alpha}_a = \mathbf{0}_p \text{ versus } K_{31}: \boldsymbol{\alpha}_i \neq \mathbf{0}_p \text{ for some } i.$$

The null hypothesis H_{31} implies that all the effects are zero.

For our high-dimensional dependent observations, we use the following Lawley–Hotelling test statistic $T_{\mathrm{LH},n}$, the likelihood ratio test statistic $T_{\mathrm{LR},n}$, and the Bartlett–Nanda–Pillai test statistic $T_{\mathrm{BNP},n}$:

$$T_{\mathrm{LH},n} := n\mathrm{tr}\hat{\mathbf{S}}_H \hat{\mathbf{S}}_E^{-1}, \; T_{\mathrm{LR},n} := -n \log |\hat{\mathbf{S}}_E| / |\hat{\mathbf{S}}_E + \hat{\mathbf{S}}_H|,$$

and

$$T_{\mathrm{BNP},n} := n\mathrm{tr}\hat{\mathbf{S}}_H (\hat{\mathbf{S}}_E + \hat{\mathbf{S}}_H)^{-1},$$

where

$$\hat{\mathbf{S}}_H := \sum_{i=1}^{a} n_i (\hat{\mathbf{Z}}_{i\cdot} - \hat{\mathbf{Z}}_{\cdot\cdot})(\hat{\mathbf{Z}}_{i\cdot} - \hat{\mathbf{Z}}_{\cdot\cdot})^\top, \text{ and } \hat{\mathbf{S}}_E := \sum_{i=1}^{a}\sum_{t=1}^{n_i} (\mathbf{Z}_{it} - \hat{\mathbf{Z}}_{i\cdot})(\mathbf{Z}_{it} - \hat{\mathbf{Z}}_{i\cdot})^\top$$

$$\text{with } \hat{\mathbf{Z}}_{i\cdot} := \frac{1}{n_i}\sum_{t=1}^{n_i} \mathbf{Z}_{it} \text{ and } \hat{\mathbf{Z}}_{\cdot\cdot} := \frac{1}{n}\sum_{i=1}^{a}\sum_{t=1}^{n_i} \mathbf{Z}_{it}.$$

Here, $\hat{\mathbf{S}}_H$ and $\hat{\mathbf{S}}_E$ are called the between-group sum of squares and the within-group sum of squares, respectively.

We derive the null asymptotic distribution of the three test statistics under the following assumptions.

Assumption 3.1 For the dimension p of $\mathbf{Z}_{it}$, the sample size of the i-th group n_i and the total sample size n satisfy

$$\frac{p^{3/2}}{\sqrt{n}} \to 0 \text{ as } n, p \to \infty \text{ and}$$
$$\frac{n_i}{n} \to \rho_i > 0 \text{ as } n \to \infty. \tag{3.2}$$

Remark 3.2 Condition (3.2) implies that the sample size of the i-th group n_i and the total sample size n of all the groups are asymptotically of the same order.

Assumption 3.2 For the p-vectors $\boldsymbol{e}_{it} = (e_{it1}, \ldots, e_{itp})^\top$ given in (3.1), there exists a non-negative integer ℓ such that

$$\sum_{s_2,\ldots,s_\ell=-\infty}^{\infty} \left\{ \left(1 + \sum_{j=2}^{\ell} |s_j|\right) |\kappa^i_{r_1,\ldots,r_\ell}(s_2, \ldots, s_\ell)| \right\} < \infty,$$

for any ℓ-tuple $r_1, \ldots, r_\ell \in \{1, \ldots, p\}$ and $i = 1, \ldots, a$. Here, $\kappa^i_{r_1,\ldots,r_\ell}(s_2, \ldots, s_\ell) := \mathrm{Cum}\{e_{i0r_1}, e_{is_2r_2}, \ldots, e_{is_\ell r_\ell}\}$, where, for a random variable $\{X_t\}$, $\mathrm{Cum}(X_1, \ldots, X_\ell)$ denotes the cumulant of order ℓ of $(X_1, \ldots, X_\ell)$ (see Brillinger, 1981, p. 19).

Remark 3.3 As per Condition (ii) at the beginning of section 3.1, $\{\boldsymbol{e}_{i\cdot}\}$, $i = 1, \ldots, a$ are mutually independent. Hence, $\mathrm{Cum}\{e_{i\cdot\cdot}, e_{j\cdot\cdot}, \cdots\} = 0$ $(i \neq j)$.

Remark 3.4 If $e_{itr_{m_1}}, \ldots, e_{itr_{m_h}}$ for some ℓ-tuple $m_1, \ldots, m_h \in \{1, \ldots, \ell\}$ are independent of $e_{itr_{m_{h+1}}}, \ldots, e_{itr_{m_\ell}}$ for the remaining $(\ell - h)$-tuple $m_{h+1}, \ldots, m_\ell \in \{1, \ldots, \ell\}$, then $\kappa^i_{r_{m_1},\ldots,r_{m_\ell}}(s_{m_1+1}, \ldots, s_{m_\ell}) = 0$ (Brillinger, 1981, p. 19). Assumption 3.2 implies that if the time points of a group of $e_{it_A r_*}$'s are well separated from the remaining time points of $e_{it_B r_*}$'s, the values of $\kappa^i_{r_1,\ldots,r_\ell}(s_2, \ldots, s_\ell)$ become small (and hence summable) (see Brillinger, 1981, p. 19). This property is natural for stochastic processes with short memory. Although some readers may believe that Assumption 3.2 is very restrictive, it is not so. Nisio (1960) studied a sequence of polynomial processes:

$$X_L(t) = \sum_{J=0}^{L} \sum_{u_1,\ldots,u_J} a_J(t - u_1, \ldots, t - u_J) W(u_1) \cdots W(u_J), \tag{3.3}$$

where the a_J's are absolutely summable, and $\{W(u)\} \overset{i.i.d.}{\sim} N(0, 1)$. Nisio (1960) showed that, when a process $\{Y(t)\}$ is strictly stationary and ergodic, there is a sequence of polynomial processes that converges to $\{Y(t)\}$ in law. The process $X_L(t)$ is called the Volterra series. Clearly, the sequence of polynomial processes (3.3) satisfies Assumption 3.2. Also, in the following (3.5), we present a very practical nonlinear time series model, which satisfies Assumption 3.2.

Assumption 3.3 The disturbance process $\{\boldsymbol{e}_{i\cdot}\}$ is an uncorrelated process, that is,

$$\boldsymbol{\Gamma}(j) = \boldsymbol{O}_p \text{ for all } j \neq 0. \tag{3.4}$$

Remark 3.5 Assumption 3.3 means that we replace independent disturbances in the classical ANOVA by dependent ones like white noise and GARCH type disturbances. It should be noted here that the condition (3.4) is of course restrictive, but includes practical nonlinear time series models such as DCC-GARCH(u, v):

$$\begin{aligned}
\boldsymbol{e}_{it} &= \boldsymbol{H}_{it}^{1/2}\boldsymbol{\eta}_{it}, \quad \boldsymbol{\eta}_{it} \overset{i.i.d.}{\sim} (\boldsymbol{0}_p, \boldsymbol{I}_p), \\
\boldsymbol{H}_{it} &= \boldsymbol{D}_{it}\boldsymbol{R}_{it}\boldsymbol{D}_{it}, \quad \boldsymbol{D}_{it} = diag\left[\sqrt{\sigma_{it1}}, \ldots, \sqrt{\sigma_{itp}}\right], \\
\boldsymbol{e}_{it} &= \begin{pmatrix} e_{it1} \\ \vdots \\ e_{itp} \end{pmatrix}, \quad \sigma_{itj} = w_j + b_j \textstyle\sum_{l=1}^{u} \sigma_{i(t-l)j} + c_j \sum_{l=1}^{v} \left\{e_{i(t-l)j}\right\}^2, \\
\boldsymbol{R}_{it} &= \left(diag\left[\boldsymbol{Q}_{it}\right]\right)^{-1/2} \boldsymbol{Q}_{it} \left(diag\left[\boldsymbol{Q}_{it}\right]\right)^{-1/2}, \\
\tilde{\boldsymbol{e}}_{it} &= \begin{pmatrix} \tilde{e}_{it1} \\ \vdots \\ \tilde{e}_{itp} \end{pmatrix}, \quad \tilde{e}_{itj} = \frac{e_{itj}}{\sqrt{\sigma_{itj}}}, \quad \boldsymbol{Q}_{it} = (1-\alpha-\beta)\tilde{\boldsymbol{Q}} + \alpha\tilde{\boldsymbol{e}}_{i(t-1)}\tilde{\boldsymbol{e}}_{i(t-1)}^{\top} + \beta\boldsymbol{Q}_{i(t-1)},
\end{aligned} \tag{3.5}$$

where $\tilde{\boldsymbol{Q}}$, the unconditional correlation matrix, is a constant positive semidefinite matrix, and $\boldsymbol{H}_{it}$'s are measurable with respect to $\boldsymbol{\eta}_{i(t-1)}, \boldsymbol{\eta}_{i(t-2)}, \cdots$ (see Engle, 2002) satisfying (3.4). By Giraitis et al. (2000, formula (2.3)), a typical component is expressed as

$$\sum_{l=0}^{\infty} \sum_{j_l < j_{l-1} < \cdots < j_1 < t} b_{t-j_1} \cdots b_{j_{l-1}-j_l} \eta_{j_1} \cdots \eta_{j_l},$$

where η_j's are i.i.d. with $E\eta_j^2 < \infty$. It is easy to observe that the DCC-GARCH(u, v) satisfies Assumption 3.2.

Remark 3.6 In what follows, our discussion for $\boldsymbol{Z}_{it}$ remains valid if we apply a linear transformation $\{\boldsymbol{\Gamma}(0)\}^{-1/2}$ to $\boldsymbol{Z}_{it}$ because the three test statistics $T_{\mathrm{LH},n}$, $T_{\mathrm{LR},n}$, and $T_{\mathrm{BNP},n}$ are invariant under linear transformation. Hence, without loss of generality, we may assume $\boldsymbol{\Gamma}(0) = \boldsymbol{I}_p$, and $\boldsymbol{\mu} = \boldsymbol{0}_p$.

3.1.1 Asymptotics of Fundamental Statistics for High-Dimensional Time Series

In this subsection, we provide lemmas and their proofs. In the following, we use the same linear transformation as in Remark 3.6. First, the stochastic expansion of $n^{-1}\hat{\boldsymbol{S}}_E$ and $\hat{\boldsymbol{S}}_H$ is given.

Lemma 3.1 *Suppose Assumptions 3.1–3.3 hold. Then, under the null hypothesis* H_{31},

$$\begin{aligned}
\frac{1}{n}\hat{\boldsymbol{S}}_E &= \boldsymbol{I}_p + O_p^U\left(\frac{1}{\sqrt{n}}\right), \\
n\hat{\boldsymbol{S}}_E^{-1} &= \boldsymbol{I}_p + O_p^U\left(\frac{1}{\sqrt{n}}\right), \quad \textit{and } \hat{\boldsymbol{S}}_H = O_p^U(1).
\end{aligned} \tag{3.6}$$

Proof (***Lemma*** 3.1) Write (3.6) as

$$\frac{1}{n}\hat{\mathbf{S}}_E = \sum_{i=1}^{a}\left(\frac{n_i}{n}\right)\frac{1}{n_i}\sum_{t=1}^{n_i}(\mathbf{Z}_{it} - \hat{\mathbf{Z}}_{i\cdot})(\mathbf{Z}_{it} - \hat{\mathbf{Z}}_{i\cdot})^\top,$$
$$= \sum_{i=1}^{a}\rho_i\frac{1}{n_i}\sum_{t=1}^{n_i}(\mathbf{Z}_{it} - \hat{\mathbf{Z}}_{i\cdot})(\mathbf{Z}_{it} - \hat{\mathbf{Z}}_{i\cdot})^\top.$$

In what follows, for each i, rewrite

$$\begin{aligned}
&\frac{1}{n_i}\sum_{t=1}^{n_i}(\mathbf{Z}_{it} - \hat{\mathbf{Z}}_{i\cdot})(\mathbf{Z}_{it} - \hat{\mathbf{Z}}_{i\cdot})^\top \\
&= \frac{1}{n_i}\sum_{t=1}^{n_i}(\mathbf{Z}_{it} - \boldsymbol{\alpha}_i + \boldsymbol{\alpha}_i - \hat{\mathbf{Z}}_{i\cdot})(\mathbf{Z}_{it} - \boldsymbol{\alpha}_i + \boldsymbol{\alpha}_i - \hat{\mathbf{Z}}_{i\cdot})^\top \\
&= \frac{1}{n_i}\sum_{t=1}^{n_i}(\mathbf{Z}_{it} - \boldsymbol{\alpha}_i)(\mathbf{Z}_{it} - \boldsymbol{\alpha}_i)^\top + (\boldsymbol{\alpha}_i - \hat{\mathbf{Z}}_{i\cdot})(\boldsymbol{\alpha}_i - \hat{\mathbf{Z}}_{i\cdot})^\top \\
&\quad + \frac{1}{n_i}\sum_{t=1}^{n_i}\left\{(\mathbf{Z}_{it} - \boldsymbol{\alpha}_i)(\boldsymbol{\alpha}_i - \hat{\mathbf{Z}}_{i\cdot})^\top + (\boldsymbol{\alpha}_i - \hat{\mathbf{Z}}_{i\cdot})(\mathbf{Z}_{it} - \boldsymbol{\alpha}_i)^\top\right\} \\
&= \boldsymbol{A} + \boldsymbol{B} + \boldsymbol{C},
\end{aligned}$$

where $\boldsymbol{A} := \{A_{jk}\}$, $\boldsymbol{B} := \{B_{jk}\}$, and $\boldsymbol{C} := \{C_{jk}\}$. We observe

$$\begin{aligned}
&\mathrm{E}\{\boldsymbol{A}\} = \boldsymbol{I}_p \quad \text{and} \\
&\mathrm{Cov}\{A_{jk}, A_{lm}\} \\
&= \frac{1}{n_i}\sum_{s=-n_i+1}^{n_i-1}\left(1 - \frac{|s|}{n_i}\right)\{\kappa^i_{j,l}(s)\kappa^i_{k,m}(s) + \kappa^i_{j,m}(s)\kappa^i_{k.l}(s) + \kappa^i_{j,k,l,m}(0,s,s)\} \\
&= O\left(n_i^{-1}\right) = O\left(n^{-1}\right) \quad \text{uniformly in } j, k, l, m \text{ by Assumption 3.3.} \qquad \square
\end{aligned}$$

Therefore, $\boldsymbol{A} = \boldsymbol{I}_p + O_p^U\left(1/\sqrt{n}\right)$. Next, we observe

$$\begin{aligned}
\mathrm{E}(\hat{\boldsymbol{Z}}_{i\cdot}) &= \boldsymbol{\alpha}_i \quad \text{and} \\
\mathrm{Cov}\{\hat{\boldsymbol{Z}}_{i\cdot}, \hat{\boldsymbol{Z}}_{i\cdot}\} &= \left(\frac{1}{n_i}\sum_{s=-n_i+1}^{n_i-1}\left(1 - \frac{|s|}{n_i}\right)\kappa^i_{j,k}(s)\right)_{1\le j,k\le p} \qquad (3.7)\\
&= O^U\left(\frac{1}{n_i}\right).
\end{aligned}$$

Thus,

$$\boldsymbol{B} = O_p^U\left(\frac{1}{n}\right).$$

Since $\boldsymbol{C}$ is a matrix of cross-product terms between $\boldsymbol{A}$ and $\boldsymbol{B}$, applying Schwarz's inequality to each component of $\boldsymbol{C}$ yields $\boldsymbol{C} = O_p^U\left(1/\sqrt{n}\right)$. Hence,

$$\frac{1}{n}\hat{\mathcal{S}}_E = \boldsymbol{I}_p + O_p^U\left(\frac{1}{\sqrt{n}}\right),$$

and

$$\left\{\frac{1}{n}\hat{\mathcal{S}}_E\right\}^{-1} = \left\{\boldsymbol{I}_p + O_p^U\left(\frac{1}{\sqrt{n}}\right)\right\}^{-1} = \{\boldsymbol{I}_p - \boldsymbol{D}_n\}^{-1} \text{ (say).}$$

It is known that

$$\{\boldsymbol{I}_p - \boldsymbol{D}_n\}^{-1} = \sum_{k=0}^{\infty} \boldsymbol{D}_n^k$$

(see Magnus and Neudecker, 1999, p. 169). From Assumption 3.1, it follows that

$$\boldsymbol{D}_n^k = O_p\left(n^{-k/2}\right)\boldsymbol{H},$$

where $\boldsymbol{H}$ is a $p \times p$-matrix and $\boldsymbol{H} = O_p^U(1)$. Then, we obtain

$$\left\{\frac{1}{n}\hat{\mathcal{S}}_E\right\}^{-1} = \boldsymbol{I}_p + O_p^U\left(\frac{1}{\sqrt{n}}\right).$$

Next, we prove that $\hat{\mathcal{S}}_H = O_p^U(1)$. To this end, let us recall

$$\hat{\boldsymbol{S}}_H = \sum_{i=1}^{a} n_i(\hat{\boldsymbol{Z}}_{i\cdot} - \hat{\boldsymbol{Z}}_{\cdot\cdot})(\hat{\boldsymbol{Z}}_{i\cdot} - \hat{\boldsymbol{Z}}_{\cdot\cdot})^\top. \tag{3.8}$$

From (3.7), we observe that $\hat{\boldsymbol{Z}}_{i\cdot} = \boldsymbol{\alpha}_i + O_p^U\left(1/\sqrt{n_i}\right)$, $\sum_{i=1}^{a}\boldsymbol{\alpha}_i = \boldsymbol{0}_p$ and, similarly, $\hat{\boldsymbol{Z}}_{\cdot\cdot} = O_p^U\left(1/\sqrt{n}\right)$. Thus, we have

$$\hat{\mathcal{S}}_H = O_p^U(1). \tag{3.9}$$

Lemma 3.2 *Suppose Assumptions 3.1–3.3 hold. Then, under the null hypothesis* H_{31},

$$\begin{aligned} n\mathrm{tr}\{\hat{\boldsymbol{S}}_H\hat{\boldsymbol{S}}_E^{-1}\} &= \mathrm{tr}\hat{\boldsymbol{S}}_H + O_p\left(\frac{p^2}{\sqrt{n}}\right) \\ &= (a-1)p + O_p\left(\frac{p^2}{\sqrt{n}}\right). \end{aligned}$$

Proof (***Lemma*** 3.2) From Lemma 3.1, we see that

$$\begin{aligned} n\mathrm{tr}\{\hat{\boldsymbol{S}}_H\hat{\boldsymbol{S}}_E^{-1}\} &= \mathrm{tr}\left[\hat{\boldsymbol{S}}_H\left(\frac{1}{n}\hat{\boldsymbol{S}}_E\right)^{-1}\right] \\ &= \mathrm{tr}\left[\hat{\boldsymbol{S}}_H\left\{\boldsymbol{I}_p + O_p^U\left(\frac{1}{\sqrt{n}}\right)\right\}\right] \\ &= \mathrm{tr}\hat{\boldsymbol{S}}_H + \mathrm{tr}\left\{\hat{\boldsymbol{S}}_H\, O_p^U\left(\frac{1}{\sqrt{n}}\right)\right\}. \end{aligned}$$

From (3.9),

$$n\mathrm{tr}\{\hat{\boldsymbol{S}}_H\hat{\boldsymbol{S}}_E^{-1}\} = \mathrm{tr}\hat{\boldsymbol{S}}_H + O_p\left(\frac{p^2}{\sqrt{n}}\right).$$

Under H_{31}, we may assume without loss of generality that $\boldsymbol{\mu} = \boldsymbol{0}_p$, $\boldsymbol{\alpha}_i = \boldsymbol{0}_p$, $i = 1, \ldots, a$. Consequently, by recalling (3.7) and (3.8), we obtain

$$E[n\mathrm{tr}\{\hat{\boldsymbol{S}}_H\hat{\boldsymbol{S}}_E^{-1}\}] = (a-1)p + O\left(\frac{p^2}{\sqrt{n}}\right),$$

which completes the proof. □

3.1.2 Test Statistics for High-Dimensional Time Series

This subsection provides the asymptotic theory for the three test statistics, $T_{\mathrm{LH},n,p}$, $T_{\mathrm{LR},n,p}$, and $T_{\mathrm{BNP},n,p}$. In the following, we use the same linear transformation as in Remark 3.6. We derive the stochastic expansion of the standardized versions $T_{\mathrm{mLH},n,p}$, $T_{\mathrm{mLR},n,p}$, and $T_{\mathrm{mBNP},n,p}$ of the three test statistics $T_{\mathrm{LH},n}$, $T_{\mathrm{LR},n}$, and $T_{\mathrm{BNP},n}$, respectively:

$$T_{\mathrm{mLH},n,p} := \frac{1}{\sqrt{2(a-1)}}\left\{\frac{n}{\sqrt{p}}\mathrm{tr}\hat{\boldsymbol{S}}_H\hat{\boldsymbol{S}}_E^{-1} - (a-1)\sqrt{p}\right\}, \tag{3.10}$$

$$T_{\mathrm{mLR},n,p} := -\frac{1}{\sqrt{2(a-1)}}\left\{\frac{n}{\sqrt{p}}\log|\hat{\boldsymbol{S}}_E|/|\hat{\boldsymbol{S}}_E + \hat{\boldsymbol{S}}_H| + (a-1)\sqrt{p}\right\}, \tag{3.11}$$

$$T_{\mathrm{mBNP},n,p} := \frac{1}{\sqrt{2(a-1)}}\left\{\frac{n}{\sqrt{p}}\mathrm{tr}\hat{\boldsymbol{S}}_H(\hat{\boldsymbol{S}}_E + \hat{\boldsymbol{S}}_H)^{-1} - (a-1)\sqrt{p}\right\}. \tag{3.12}$$

Theorem 3.1 *Suppose Assumptions 3.1–3.3 hold. Then, under the null hypothesis H_{31}, we have the stochastic expansion: for any $i \in \{\mathrm{mLH}, \mathrm{mLR}, \mathrm{mBNP}\}$,*

$$T_{i,n,p} = \frac{1}{\sqrt{2(a-1)}}\left\{\frac{1}{\sqrt{p}}\mathrm{tr}\hat{\mathcal{S}}_H - (a-1)\sqrt{p}\right\} + o_p(1).$$

Proof (***Theorem*** 3.1) From Lemma 3.2, it is immediately evident that

$$T_{\text{mLH},n,p} = \frac{1}{\sqrt{2(a-1)}}\left\{\frac{1}{\sqrt{p}}\text{tr}\hat{\mathcal{S}}_H - (a-1)\sqrt{p}\right\} + o_p\left(\frac{p^{3/2}}{\sqrt{n}}\right),$$

whose error term becomes $o_p(1)$ by Assumption 3.1. For $T_{\text{mLR},n,p}$, first, note that

$$d\log|\boldsymbol{F}| = \text{tr}(\boldsymbol{F}^{-1}d\boldsymbol{F})$$

(e.g., Magnus and Neudecker, 1999). Then, a modification of Proposition 6.1.5 of Brockwell and Davis (1991), and Lemmas 3.1 and 3.2 show

$$T_{\text{mLR},n,p} = \frac{1}{\sqrt{2(a-1)}}\left\{\frac{1}{\sqrt{p}}\text{tr}\hat{\mathcal{S}}_H - (a-1)\sqrt{p}\right\} + o_p(1).$$

For $T_{\text{mBNP},n,p}$, in view of Lemmas 3.1 and 3.2, we obtain

$$\begin{aligned} n\text{tr}\{\hat{\boldsymbol{S}}_H(\hat{\boldsymbol{S}}_E + \hat{\boldsymbol{S}}_H)^{-1}\} &= \text{tr}\left[\hat{\boldsymbol{S}}_H\left(\frac{1}{n}\hat{\boldsymbol{S}}_E + \frac{1}{n}\hat{\boldsymbol{S}}_H\right)^{-1}\right] \\ &= \text{tr}\left\{\hat{\boldsymbol{S}}_H\left(\boldsymbol{I}_p + O_p^U\left(\frac{1}{\sqrt{n}}\right)\right)^{-1}\right\} \\ &= \text{tr}\left\{\hat{\boldsymbol{S}}_H\left(\boldsymbol{I}_p + O_p^U\left(\frac{1}{\sqrt{n}}\right)\right)\right\}, \end{aligned}$$

which leads to

$$T_{\text{mBNP},n,p} = \frac{1}{\sqrt{2(a-1)}}\left\{\frac{1}{\sqrt{p}}\text{tr}\hat{\mathcal{S}}_H - (a-1)\sqrt{p}\right\} + o_p(1).$$

We obtained the desired result.

By Lemmas 3.1 and 3.2 and Theorem 3.1, the asymptotic distributions of $T_{\text{mLH},n,p}$, $T_{\text{mLR},n,p}$, and $T_{\text{mBNP},n,p}$ are derived in the following theorem.

Theorem 3.2 *Suppose Assumptions 3.1–3.3 hold. Then, under the null hypothesis* H_{31}*, for any* $i \in \{\text{mLH, mLR, mBNP}\}$, $T_{i,n,p}$ *converges in distribution to the standard normal.*

Proof (***Theorem*** 3.2) From Lemma 3.2, it is evident that

$$\frac{1}{\sqrt{p}}E\{\text{tr}\hat{\mathcal{S}}_H\} = (a-1)\sqrt{p} + O\left(\frac{p^{3/2}}{\sqrt{n}}\right).$$

Furthermore, by (3.7) and (3.9), it is easy to show

$$\mathrm{Cum}\left\{\frac{1}{\sqrt{p}}\mathrm{tr}\hat{\mathcal{S}}_H, \frac{1}{\sqrt{p}}\mathrm{tr}\hat{\mathcal{S}}_H\right\} = 2(a-1) + o(1).$$

Because $p^{-1/2}\mathrm{tr}\hat{\mathcal{S}}_H$ is a finite linear combination of

$$\begin{aligned}
A_{ii} &:= \frac{1}{\sqrt{p}} n_i (\hat{\mathbf{Z}}_{i\cdot} - \mathbf{Z}_{\cdot\cdot})^\top (\hat{\mathbf{Z}}_{i\cdot} - \mathbf{Z}_{\cdot\cdot}) \\
&= \frac{1}{\sqrt{p}}\frac{1}{n_i}\sum_{l_1=1}^{p}\sum_{t_1=1}^{n_i} e_{it_1l_1}\sum_{s_1=1}^{n_i} e_{is_1l_1} - \frac{2}{\sqrt{p}}\frac{1}{n}\sum_{l_2=1}^{p}\sum_{t_2=1}^{n_i} e_{it_2l_2}\sum_{j=1}^{a}\sum_{s_2=1}^{n_j} e_{js_2l_2} \\
&\quad + \frac{1}{\sqrt{p}}\frac{n_i}{n^2}\sum_{l_3=1}^{p}\sum_{j_1=1}^{a}\sum_{t_3=1}^{n_{j_1}} e_{j_1t_3l_3}\sum_{j_2=1}^{a}\sum_{s_3=1}^{n_{j_2}} e_{j_2s_3l_3} \\
&\approx \frac{1}{\sqrt{p}}\frac{1}{n}\sum_{l_1=1}^{p}\sum_{t_1=1}^{n_i}\sum_{s_1=1}^{n_i} e_{it_1l_1}e_{is_1l_1} - \frac{2}{\sqrt{p}}\frac{1}{n}\sum_{l_2=1}^{p}\sum_{j=1}^{a}\sum_{t_2=1}^{n_i}\sum_{s_2=1}^{n_j} e_{it_2l_2}e_{js_2l_2} \\
&\quad + \frac{1}{\sqrt{p}}\frac{\rho_i}{n}\sum_{l_3=1}^{p}\sum_{j_1=1}^{a}\sum_{j_2=1}^{a}\sum_{t_3=1}^{n_{j_1}}\sum_{s_3=1}^{n_{j_2}} e_{j_1t_3l_3}e_{j_2s_3l_3} \\
&:= B_{ii} + C_{ii} + D_{ii},
\end{aligned}$$

we can evaluate $\mathrm{cum}^{(J)}\left\{p^{-1/2}\mathrm{tr}\hat{\mathcal{S}}_H, \ldots, p^{-1/2}\mathrm{tr}\hat{\mathcal{S}}_H\right\}$ by using the properties of the cumulants (see Brillinger, 1981, p. 19).

Next, we show that the J-th order ($J \geq 3$) cumulants of A_{ii} tend to zero; since C_{ii} and D_{ii} can be treated similarly, it is sufficient to focus on the term B_{ii}. Hence, we only show that the J-th order ($J \geq 3$) cumulants of B_{ii} converge to zero. In fact,

$$\begin{aligned}
\mathrm{Cum}\{B_{i_1i_1}, B_{i_2i_2}, \ldots, B_{i_Ji_J}\} = \Bigg[p^{-\frac{J}{2}}n^{-J}\sum_{l_1=1}^{p}\cdots\sum_{l_J=1}^{p}\sum_{t_1=1}^{n_{i_1}}\cdots\sum_{t_J=1}^{n_{i_J}}\sum_{s_1=1}^{n_{i_1}}\cdots\sum_{s_J=1}^{n_{i_J}} \\
\mathrm{Cum}\left\{e_{i_1t_1l_1}e_{i_1s_1l_1}, \ldots, e_{i_Jt_Jl_J}e_{i_Js_Jl_J}\right\}\Bigg]. \qquad (3.13)
\end{aligned}$$

Brillinger (1981) shows that the J-th order cumulant Cum{ , . . . , } is expressed as the indecomposable sum of products of $\mathrm{Cum}\{e_{itl} : (t, l) \in \nu\}$. Hence, a typical main-order term is

$$\mathrm{Cum}\left\{e_{i_1s_1l_1}, e_{i_2t_2l_2}\right\}\mathrm{Cum}\left\{e_{i_2s_2l_2}, e_{i_3t_3l_3}\right\}\cdots\mathrm{Cum}\left\{e_{i_Js_Jl_J}, e_{i_1t_1l_1}\right\}. \qquad (3.14)$$

From Remark 3.3,

$$\mathrm{Cum}\left\{e_{is_1l_1}, e_{it_2l_2}\right\}\mathrm{Cum}\left\{e_{is_2l_2}, e_{it_3l_3}\right\}\cdots\mathrm{Cum}\left\{e_{is_Jl_J}, e_{it_1l_1}\right\} \tag{3.15}$$

in (3.14) remains. Therefore, the main-order term of (3.13) for the typical cumulant (3.15) is

$$p^{-\frac{J}{2}}n^{-J}\sum_{l_1=1}^{p}\cdots\sum_{l_J=1}^{p}\sum_{t_1=1}^{n_{i_1}}\cdots\sum_{t_J=1}^{n_{i_J}}\sum_{s_1=1}^{n_{j_1}}\cdots\sum_{s_J=1}^{n_{j_J}}\kappa^i_{l_1,l_2}(t_2-s_1)\kappa^i_{l_2,l_3}(t_3-s_2)\cdots\kappa^i_{l_J,l_1}(t_1-s_J)$$

$$= p^{-\frac{J}{2}}\sum_{l=1}^{p}\sum_{r_1}\cdots\sum_{r_J}\kappa^i_{l,l}(r_1)\kappa^i_{l,l}(r_2)\cdots\kappa^i_{l,l}(r_J) \quad (\text{ by Assumption 3.3 and } \mathbf{\Gamma}(0)=\boldsymbol{I}_p)$$

$$= O\left(p^{-\frac{J}{2}+1}\right) \quad (\text{by Assumption 3.2}).$$

Finally, under Assumption 3.1, we use a method by Brillinger (1981), which shows that all the cumulants of order greater than 2 converge to zero. Thus, the characteristic functions of the standardized test statistics converge to $\exp\{-t^2/2\}$ (see Theorem 2.3.1 of Brillinger, 1981, p. 19; e.g., the proof of Theorem 5.10.1 of Brillinger, 1981, pp. 417–18). Hence, the asymptotic normality of $T_{\mathrm{ALL},n,p}$ is proved. □

References

Brillinger, D. R. (1981). *Time series: Data analysis and theory*. San Francisco: Holden-Day.

Brockwell, P. J., & Davis, R. A. (1991). *Time series: Theory and methods*. New York: Springer.

Engle, R. (2002). Dynamic conditional correlation: A simple class of multivariate generalized autoregressive conditional heteroskedasticity models. *Journal of Business & Economic Statistics, 20*, 339–350.

Giraitis, L., Kokoszka, P., & Leipus, R. (2000). Stationary ARCH models: Dependence structure and central limit theorem. *Economic Theory, 16*, 3–22.

Magnus, J. R., & Neudecker, H. (1999). *Matrix differential calculus with applications in statistics and econometrics*. New York: Wiley.

Nagahata, H., & Taniguchi, M. (2018). Analysis of variance for high-dimensional time series. *Statistical Inference for Stochastic Processes, 21*, 455–468.

Nisio, M. (1960). On polynomial approximation for strictly stationary processes. *The Journal of the Mathematical Society of Japan, 12*, 207–226.

Chapter 4
One-Way Fixed and Random Effect Models for Correlated Groups

The independence between groups was imposed in Chapter 2 but this assumption is not fulfilled by some real data, e.g., stock prices. This chapter extends one-way fixed models with time-dependent errors and independent groups dealt with in Chapter 2 to one-way fixed and random models with time-dependent errors and correlated groups. We propose a test for the existence of fixed effects in Section 4.1 and a test for the existence of random effects in Section 4.2. We show that the tests have asymptotically size φ and are consistent. The nontrivial power under the local alternative is elucidated. In what follows, we develop our discussion based on the results by Goto et al. (2023).

4.1 Test for the Existence of Fixed Effects

In this section, we introduce one-way fixed effect models with time-dependent errors and correlated groups and consider a test for the existence of fixed effects.

Let us consider the one-way fixed effect models with time-dependent errors and correlated groups

$$\boldsymbol{Z}_{it} = \boldsymbol{\mu} + \boldsymbol{\alpha}_i + \boldsymbol{e}_{it} \quad \text{for } i = 1, \ldots, a;\ t = 1, \ldots, n_i, \tag{4.1}$$

where $\boldsymbol{Z}_{it} = (z_{it1}, \ldots, z_{itp})^\top$ is the t-th p-dimensional observation from the i-th group, $\boldsymbol{\mu} = (\mu_1, \ldots, \mu_p)^\top$ is a p-dimensional general mean, $\boldsymbol{\alpha}_i = (\alpha_{i1}, \ldots, \alpha_{ip})^\top$ is a fixed effect, and $\boldsymbol{e}_{it} = (e_{it1}, \ldots, e_{itp})^\top$ is a p-dimensional time series of the i-th group at time t. The number of groups, the length of the time series from the i-th group, and the dimension of the time series from each group are denoted by a, n_i, and p.

Suppose that

(i) the observed stretch $\{\boldsymbol{Z}_{it}; i = 1, \ldots, a, t = 1, \ldots, n_i\}$ is available,
(ii) the ap-dimensional time series $(\boldsymbol{e}_{1t}^\top, \ldots, \boldsymbol{e}_{at}^\top)^\top$ is a centered stationary process, which has an ap-by-ap spectral density matrix $\boldsymbol{f}(\lambda) = (\boldsymbol{f}_{ij}(\lambda))_{i,j=1,\ldots,a}$ for

Y. Goto et al., *ANOVA with Dependent Errors*, JSS Research Series in Statistics,
https://doi.org/10.1007/978-981-99-4172-8_4

$\lambda \in [-\pi, \pi]$, where $\boldsymbol{f}_{ii}(\lambda)$ is a p-by-p spectrum of $\{\boldsymbol{e}_{it}\}$ and $\boldsymbol{f}_{ij}(\lambda)$ is a p-by-p cross spectrum of $\{\boldsymbol{e}_{it}\}$ and $\{\boldsymbol{e}_{jt}\}$ for $i \neq j$, and

(iii) there exists a constant $\rho_i \in (0, 1)$ such that $n_i = \rho_i n$ with $n = \sum_{i=1}^{a} n_i$ for any $i \in \{1, \ldots, a\}$.

Moreover, without loss of generality, we assume that the fixed effect $\boldsymbol{\alpha}_i$ satisfies $\sum_{i=1}^{a} \boldsymbol{\alpha}_i = \mathbf{0}_p$, where $\mathbf{0}_p$ is a p-dimensional vector whose entries are all zero. Actually, when $\sum_{i=1}^{a} \boldsymbol{\alpha}_i \neq \mathbf{0}_p$, we can replace μ and $\boldsymbol{\alpha}_i$ as $\mu + \sum_{i=1}^{a} \boldsymbol{\alpha}_i$ and $\boldsymbol{\alpha}_i - \sum_{i=1}^{a} \boldsymbol{\alpha}_i$, respectively.

Remark 4.1 The above Condition (ii) enables us to deal not only with the case with within-group correlation but also with the case with between-groups correlation. Condition (iii) covers the case of equal sample size for each group.

Remark 4.2 The case that more than one time series from each group is observed can be dealt with by the model (4.1) as follows: replace p as pq, where q is the number of time series in each group, and redefine

$$\begin{aligned}
\mathbf{Z}_{it} &= (z_{it11}, \ldots, z_{it1q}, z_{it21}, \ldots, z_{itp1}, \ldots z_{itpq})^\top, \quad \boldsymbol{\mu} = (\mu_1 \mathbf{1}_q^\top, \ldots, \mu_p \mathbf{1}_q^\top)^\top, \\
\boldsymbol{\alpha}_i &= (\alpha_{i1} \mathbf{1}_q^\top, \ldots, \alpha_{ip} \mathbf{1}_q^\top)^\top, \quad \boldsymbol{e}_{it} = (e_{it11}, \ldots, e_{it1q}, e_{it12}, \ldots, e_{itp1}, \ldots, e_{itpq})^\top,
\end{aligned}$$

where $\mathbf{1}_q$ is a q-dimensional vector whose entries are all one. For more general cases, p and q can depend on i.

Next, we formulate the test for the existence of fixed effects. The null hypothesis H_{41} and the alternative hypothesis K_{41} are defined as

$$H_{41} : \boldsymbol{\alpha}_1 = \cdots = \boldsymbol{\alpha}_a \quad \text{vs} \quad K_{41} : \boldsymbol{\alpha}_i \neq \mathbf{0}_p \ \text{ for some } i. \tag{4.2}$$

The null hypothesis is equivalent to $\boldsymbol{\alpha}_i = \mathbf{0}_p$ for all $i \in \{1, \ldots, a\}$ by the condition $\sum_{i=1}^{a} \boldsymbol{\alpha}_i = \mathbf{0}_p$.

4.1.1 Classical Test Statistics

In the context of ANOVA, the sum of squares quantities play an important role. A simple algebra gives

$$\sum_{i=1}^{a} \sum_{t=1}^{n_i} (\mathbf{Z}_{it} - \mathbf{Z}_{..})^\top (\mathbf{Z}_{it} - \mathbf{Z}_{..}) \tag{4.3}$$

$$= \sum_{i=1}^{a} n_i (\mathbf{Z}_{i.} - \mathbf{Z}_{..})^\top (\mathbf{Z}_{i.} - \mathbf{Z}_{..}) + \sum_{i=1}^{a} \sum_{t=1}^{n_i} (\mathbf{Z}_{it} - \mathbf{Z}_{i.})^\top (\mathbf{Z}_{it} - \mathbf{Z}_{i.}), \tag{4.4}$$

where $\boldsymbol{Z}_{i.} := \sum_{t=1}^{n_i} \boldsymbol{Z}_{it}/n_i$ and $\boldsymbol{Z}_{..} := \sum_{i=1}^{a} \sum_{t=1}^{n_i} \boldsymbol{Z}_{it}/(n_i a)$. This equation shows that the total sum of squares (4.3) is decomposed into the between-groups sum of squares (the first term of (4.4)) and the within-group sum of squares (the second term of (4.4)).

The multivariate version of the classical F-statistic is defined under the homogeneity of variances for disturbances among groups and i.i.d. assumption for disturbances, the independence of groups, and the balanced design ($n_1 = \cdots = n_a$), as

$$T_{\mathrm{iid},n} := n_0 \sum_{i=1}^{a} (\boldsymbol{Z}_{i.} - \boldsymbol{Z}_{..})^\top \hat{V}_{\mathrm{iid},n}^{-1} (\boldsymbol{Z}_{i.} - \boldsymbol{Z}_{..}), \tag{4.5}$$

where

$$\hat{V}_{\mathrm{iid},n} := \frac{1}{a} \sum_{i=1}^{a} \left(\frac{1}{n_0} \sum_{t=1}^{n_0} (\boldsymbol{Z}_{it} - \boldsymbol{Z}_{i.})^\top (\boldsymbol{Z}_{it} - \boldsymbol{Z}_{i.}) \right)$$

with $n_0 := n_1 = \cdots = n_a$. This test statistic takes the form of the ratio of the between-groups sum of squares and within-groups sum of squares. The statistic (4.5) can be rewritten in the following form:

$$T_{\mathrm{iid},n} = n_0 \begin{pmatrix} \hat{V}_{\mathrm{iid},n}^{-1/2} (\boldsymbol{\alpha}_1 + \boldsymbol{e}_{1.}) \\ \vdots \\ \hat{V}_{\mathrm{iid},n}^{-1/2} (\boldsymbol{\alpha}_a + \boldsymbol{e}_{a.}) \end{pmatrix}^\top \left\{ (\boldsymbol{I}_a - \boldsymbol{J}_a/a) \otimes \boldsymbol{I}_p \right\} \begin{pmatrix} \hat{V}_{\mathrm{iid},n}^{-1/2} (\boldsymbol{\alpha}_1 + \boldsymbol{e}_{1.}) \\ \vdots \\ \hat{V}_{\mathrm{iid},n}^{-1/2} (\boldsymbol{\alpha}_a + \boldsymbol{e}_{a.}) \end{pmatrix},$$

where $\boldsymbol{I}_a$ and $\boldsymbol{J}_a$ denote the identity matrix of size a and the a-by-a matrix whose entries are all one, respectively. Under the the null hypothesis, we can show that

$$\sqrt{n_0} \begin{pmatrix} \hat{V}_{\mathrm{iid},n}^{-1/2} \boldsymbol{e}_{1.} \\ \vdots \\ \hat{V}_{\mathrm{iid},n}^{-1/2} \boldsymbol{e}_{a.} \end{pmatrix} \Rightarrow N(\boldsymbol{0}, \boldsymbol{I}_{ap}) \quad \text{as } n_0 \to \infty,$$

provided regularity conditions. Hence, $T_{\mathrm{iid},n}$ converges in distribution to the chi-square distribution with $(a-1)p$ degrees of freedom since $(\boldsymbol{I}_a - \boldsymbol{J}_a/a) \otimes \boldsymbol{I}_p$ is idempotent and rank $\left\{ (\boldsymbol{I}_a - \boldsymbol{J}_a/a) \otimes \boldsymbol{I}_p \right\} = (a-1)p$. The test statistic (4.5) cannot be applied to time series data since the time series structure is not taken into account. A natural remedy for this problem is replacing $\hat{V}_{\mathrm{iid},n}$ with

$$\hat{V}_{\mathrm{ts},n} := \left(\frac{2\pi}{a} \sum_{i=1}^{a} \hat{\boldsymbol{f}}_{ii}(0) \right),$$

where $\hat{\boldsymbol{f}}_{ii}(0)$ is a consistent estimator of $\boldsymbol{f}_{ii}(0)$ for $i = 1, \ldots, a$, i.e., under the stationarity and the homogeneity of the spectra for disturbances, the independence of groups, and the balanced design ($n_1 = \cdots = n_a$). The test statistic is defined as

$$T_{\mathrm{ts},n} := n_0 \sum_{i=1}^{a} (\boldsymbol{Z}_{i.} - \boldsymbol{Z}_{..})^\top \hat{V}_{\mathrm{ts},n}^{-1} (\boldsymbol{Z}_{i.} - \boldsymbol{Z}_{..}). \tag{4.6}$$

This test statistic, which is considered in Chapter 2, can be interpreted as being standardized by the spectrum for each group. The asymptotic null distribution of (4.6) is the chi-square distribution with $(a-1)p$ degrees of freedom under the regularity conditions, which can be proved by the same argument as in the derivation of the asymptotic null distribution for $T_{\mathrm{iid},n}$.

In the case of correlated groups, $T_{\mathrm{iid},n}$ is not asymptotically distribution-free since

$$\sqrt{n_0} \begin{pmatrix} \hat{V}_{\mathrm{ts},n}^{-1/2} \boldsymbol{e}_{1.} \\ \vdots \\ \hat{V}_{\mathrm{ts},n}^{-1/2} \boldsymbol{e}_{a.} \end{pmatrix} \Rightarrow N\left(\boldsymbol{0}, \left(\frac{2\pi}{a} \sum_{i=1}^{a} \boldsymbol{f}_{ii}(0)\right)^{-1} \begin{pmatrix} \boldsymbol{f}_{11}(0) & \cdots & \boldsymbol{f}_{1a}(0) \\ \vdots & \ddots & \vdots \\ \boldsymbol{f}_{a1}(0) & \cdots & \boldsymbol{f}_{aa}(0) \end{pmatrix}\right)$$

as $n_0 \to \infty$, under the balanced design. In the next paragraph, we consider the test statistic endowed with the asymptotically distribution-freeness under the correlated groups.

4.1.2 A Test Statistic for Correlated Groups

For correlated groups, the test for the hypothesis (4.2) is developed by Goto et al. (2023). The key idea is to consider the standardized statistic not only within group but also between groups.

We make the following assumption to ensure the subsequent Lemma 4.1 to hold the following.

Assumption 4.1 For all $\ell \in \mathbb{N}$, $(k_1, \ldots, k_\ell) \in \{1, \ldots, a\}^\ell$, and $(r_1, \ldots, r_\ell) \in \{1, \ldots, p\}^\ell$,

$$\sum_{s_2, \ldots, s_\ell = -\infty}^{\infty} \left\{ \left(1 + \sum_{j=2}^{\ell} |s_j|\right) \left|\kappa_{r_1 \cdots r_\ell}^{k_1 \cdots k_\ell}(s_2, \ldots, s_\ell)\right| \right\} < \infty,$$

where $\kappa_{r_1 \cdots r_\ell}^{k_1 \cdots k_\ell}(s_2, \ldots, s_\ell) := \mathrm{Cum}\{e_{k_1 0 r_1}, e_{k_2 s_2 r_2}, \ldots, e_{k_\ell s_\ell r_\ell}\}$ and, for a random variable $\{X_t\}$, $\mathrm{Cum}(X_1, \cdots, X_\ell)$ denotes the cumulant of order ℓ of $(X_1, \cdots, X_\ell)$ (see Brillinger 1981, p.19).

Lemma 4.1 *Suppose Assumption 4.1 holds. Under H_{41} defined in* (4.2)*, we have*

$$\sqrt{n}(\boldsymbol{Z}_{1.}^\top - \boldsymbol{Z}_{..}^\top, \ldots, \boldsymbol{Z}_{a.}^\top - \boldsymbol{Z}_{..}^\top)^\top \Rightarrow N(\boldsymbol{0}, \boldsymbol{V}) \quad as \min_{i=1,\ldots,a} n_i \to \infty,$$

where $\boldsymbol{V} = (\boldsymbol{V}_{ij})_{i,j=1\ldots,a}$ *with*

$$\boldsymbol{V}_{ij} := \frac{2\pi \min\{\rho_i, \rho_j\}}{\rho_i \rho_j} \boldsymbol{f}_{ij}(0) - \frac{2\pi}{a} \sum_{s=1}^{a} \left\{ \frac{\min\{\rho_s, \rho_j\}}{\rho_s \rho_j} \boldsymbol{f}_{sj}(0) + \frac{\min\{\rho_i, \rho_s\}}{\rho_i \rho_s} \boldsymbol{f}_{is}(0) \right\}$$
$$+ \frac{2\pi}{a^2} \sum_{s,k=1}^{a} \frac{\min\{\rho_s, \rho_k\}}{\rho_s \rho_k} \boldsymbol{f}_{sk}(0).$$

Proof (Lemma 4.1)

First, we have, for any $i \in \{1, \ldots, a\}$, $\sqrt{n}\mathrm{E}(\mathbf{Z}_{i.} - \mathbf{Z}_{..}) = \mathbf{0}$.

Second, we observe for $i, j \in \{1, \ldots, a\}$ and $n_j > n_i$,

$$n\mathrm{Cov}(\mathbf{Z}_{i.}, \mathbf{Z}_{j.}) = \frac{1}{\rho_i \rho_j n} \sum_{t_i=1}^{n_i} \sum_{t_j=1}^{n_j} \phi(n_i, n_j, s) \boldsymbol{\Gamma}_{ij}(t_i - t_j)$$
$$= \frac{1}{\rho_i \rho_j n} \sum_{s=-\infty}^{\infty} \phi(n_i, n_j, s) \boldsymbol{\Gamma}_{ij}(s),$$

where

$$\phi(n_i, n_j, s) = \begin{cases} 0 & \text{if } -\infty < s \le -n_j, \\ n_j + s & \text{if } -n_j \le s \le n_i - n_j, \\ n_i & \text{if } n_i - n_j \le s \le 0, \\ n_i - s & \text{if } 0 \le s \le n_i, \\ 0 & \text{if } n_i \le s < \infty. \end{cases}$$

By Assumption 4.1, we know that

$$\left| \frac{1}{\rho_i \rho_j n} \sum_{s=-n_j+1}^{n_i - n_j} (n_j + s) \boldsymbol{\Gamma}_{ij}(s) \right| \le \frac{\rho_i + 1}{\rho_i \rho_j} \sum_{s=-n_j+1}^{n(\rho_i - \rho_j)} \left| \boldsymbol{\Gamma}_{ij}(s) \right| \to 0 \quad \text{as } n_i \to \infty,$$

$$\frac{1}{\rho_i \rho_j n} \sum_{s=n_i - n_j + 1}^{0} n_i \boldsymbol{\Gamma}_{ij}(s) = \frac{1}{\rho_j} \sum_{s=n(\rho_i - \rho_j)+1}^{0} \boldsymbol{\Gamma}_{ij}(s) \to \frac{1}{\rho_j} \sum_{s=-\infty}^{0} \boldsymbol{\Gamma}_{ij}(s) \quad \text{as } n_i \to \infty,$$

and

$$\frac{1}{\rho_i \rho_j n} \sum_{s=1}^{n_i} (n_i - s) \boldsymbol{\Gamma}_{ij}(s) = \frac{1}{\rho_j} \sum_{s=1}^{n_i} \boldsymbol{\Gamma}_{ij}(s) - \frac{1}{\rho_i \rho_j} \sum_{s=1}^{n_i} \frac{s}{n} \boldsymbol{\Gamma}_{ij}(s)$$
$$\to \frac{1}{\rho_j} \sum_{s=1}^{\infty} \boldsymbol{\Gamma}_{ij}(s) \quad \text{as } n_i \to \infty,$$

which shows that $n\mathrm{Cov}(\boldsymbol{Z}_{i.}, \boldsymbol{Z}_{j.}) \to 2\pi \boldsymbol{f}_{ij}(0)/\rho_j$ as $n_i \to \infty$. Similarly, we can show that, for any $n_j \le n_i$, $n\mathrm{Cov}(\boldsymbol{Z}_{i.}, \boldsymbol{Z}_{j.}) \to 2\pi \min\{\rho_i, \rho_j\}\boldsymbol{f}_{ij}(0)/(\rho_i\rho_j)$ as $n_j \to \infty$. Along the same line of the above arguments, $n\mathrm{Cov}(\boldsymbol{Z}_{i.} - \boldsymbol{Z}_{..}, \boldsymbol{Z}_{j.} - \boldsymbol{Z}_{..})$ converges to $\boldsymbol{V}_{ij}$ as $\min_{i=1,\ldots,a} n_i \to \infty$.

It remains only to show, for any $\ell \ge 3$, $(k_1, \ldots, k_\ell) \in \{1, \ldots, a\}^\ell$, and $(r_1, \ldots, r_\ell) \in \{1, \ldots, p\}^\ell$ that

$$n^{\ell/2}\mathrm{Cum}\{z_{k_1.r_1}, \ldots, z_{k_\ell.r_\ell}\} = O(n^{-\ell/2+1}),$$

where $z_{k_i.r_i} = \sum_{t_i=1}^{n_{k_i}} z_{k_it_ir_i}/n_{k_i}$. Assumption 4.1 yields

$$\begin{aligned}
&|n^{\ell/2}\mathrm{Cum}\{z_{k_1.r_1}, \ldots, z_{k_\ell.r_\ell}\}| \\
\le &\frac{n^{-\ell/2}}{\rho_{k_1}\cdots\rho_{k_\ell}} \sum_{t_1=1}^{n_{k_1}} \cdots \sum_{t_\ell=1}^{n_{k_\ell}} |\mathrm{Cum}\{z_{k_1t_1r_1}, \ldots, z_{k_\ell.r_\ell}\}| \\
\le &\frac{n^{-\ell/2}}{\rho_{k_1}\cdots\rho_{k_\ell}} \sum_{t_1=1}^{n_{k_1}} \sum_{s_2=1-t_1}^{n_{k_2}-t_1} \cdots \sum_{s_\ell=1-t_1}^{n_{k_\ell}-t_1} |\kappa_{r_1\cdots r_\ell}^{k_1\cdots k_\ell}(s_2, \ldots, s_\ell)| \\
\le &\frac{n^{-\ell/2}}{\rho_{k_1}\cdots\rho_{k_\ell}} \sum_{t_1=1}^{n_{k_1}} \sum_{s_2=-\infty}^{\infty} \cdots \sum_{s_\ell=-\infty}^{\infty} |\kappa_{r_1\cdots r_\ell}^{k_1\cdots k_\ell}(s_2, \ldots, s_\ell)| \\
= &\frac{n^{-\ell/2+1}}{\rho_{k_1}^2\cdots\rho_{k_\ell}} \sum_{s_2=-\infty}^{\infty} \cdots \sum_{s_\ell=-\infty}^{\infty} |\kappa_{r_1\cdots r_\ell}^{k_1\cdots k_\ell}(s_2, \ldots, s_\ell)| \\
= &O(n^{-\ell/2+1}).
\end{aligned}$$

Remark 4.3 Assumption 4.1 is often considered for dependent observations (see Brillinger, 1981, p.26), and it can be sufficiently relaxed by Theorem 2.1 of Hosoya and Taniguchi (1982).

By Lemma 4.1, we expect that

$$n(\boldsymbol{Z}_{1.}^\top - \boldsymbol{Z}_{..}^\top, \ldots, \boldsymbol{Z}_{a.}^\top - \boldsymbol{Z}_{..}^\top)^\top \boldsymbol{V}^{-1}(\boldsymbol{Z}_{1.}^\top - \boldsymbol{Z}_{..}^\top, \ldots, \boldsymbol{Z}_{a.}^\top - \boldsymbol{Z}_{..}^\top)$$

converges in distribution to the chi-square distribution with ap degrees of freedom if $\boldsymbol{V}$ is non-singular. Unfortunately, $\boldsymbol{V}$ is a singular matrix since $\sum_{i=1}^a \boldsymbol{V}_{ij} = \boldsymbol{O}_p$ for any $j \in \{1, \ldots, a\}$, where $\boldsymbol{O}_p$ is a p-by-p matrix whose entries are all zero. In order to overcome this inconvenience, we use the following lemma (Rao and Mitra, 1971, Theorem 9.2.3, p.173).

Lemma 4.2 *If $\mathfrak{Z}$ follows $N(\mathfrak{m}, \mathfrak{V})$, then it holds that $\mathfrak{Z}^\top\mathfrak{V}^-\mathfrak{Z}$ follows the noncentral chi-square distribution with* $\mathrm{rank}(\mathfrak{V})$ *degrees of freedom and noncentrality parameter $\mathfrak{m}^\top\mathfrak{V}^-\mathfrak{m}$, where $\mathfrak{V}^-$ denotes the Moore–Penrose inverse of $\mathfrak{V}$.*

This lemma allows us to obtain the asymptotic distribution by employing the Moore–Penrose inverse of $\boldsymbol{V}$ instead of the usual inverse. Thus, we consider the following quantity:

$$n(\boldsymbol{Z}_{1.}^\top - \boldsymbol{Z}_{..}^\top, \ldots, \boldsymbol{Z}_{a.}^\top - \boldsymbol{Z}_{..}^\top)^\top \boldsymbol{V}^-(\boldsymbol{Z}_{1.}^\top - \boldsymbol{Z}_{..}^\top, \ldots, \boldsymbol{Z}_{a.}^\top - \boldsymbol{Z}_{..}^\top),$$

where $\boldsymbol{V}^-$ denotes the Moore–Penrose inverse of $\boldsymbol{V}$.

Next, we estimate $\boldsymbol{V}^-$, which is a function of the spectrum $\boldsymbol{f}(0)$. A natural estimator of $\boldsymbol{V}$ can be obtained by replacing the spectrum $\boldsymbol{f}(0)$ in $\boldsymbol{V}$ with a nonparametric spectral density matrix estimator $\hat{\boldsymbol{f}}_n(0) = (\hat{\boldsymbol{f}}_{ij}(0))_{i,j=1,\ldots,a}$ of $\boldsymbol{f}(0)$, where $\hat{\boldsymbol{f}}_{ij}(0)$ is a p-by-p matrix, defined by $\hat{\boldsymbol{V}}_{\mathrm{GALT},n} = (\hat{\boldsymbol{V}}_{ij})_{i,j=1\ldots,a}$, where

$$\hat{\boldsymbol{V}}_{ij} := \frac{2\pi \min\{\rho_i, \rho_j\}}{\rho_i \rho_j} \hat{\boldsymbol{f}}_{ij}(0) - \frac{2\pi}{a} \sum_{s=1}^{a} \left\{ \frac{\min\{\rho_s, \rho_j\}}{\rho_s \rho_j} \hat{\boldsymbol{f}}_{sj}(0) + \frac{\min\{\rho_i, \rho_s\}}{\rho_i \rho_s} \hat{\boldsymbol{f}}_{is}(0) \right\}$$
$$+ \frac{2\pi}{a^2} \sum_{s,k=1}^{a} \frac{\min\{\rho_s, \rho_k\}}{\rho_s \rho_k} \hat{\boldsymbol{f}}_{sk}(0).$$

The question which naturally arises now is whether $\hat{\boldsymbol{V}}_{\mathrm{GALT},n}^-$ is a consistent estimator of $\boldsymbol{V}^-$. We make the following two assumptions in order to the estimate $\boldsymbol{V}^-$:

Assumption 4.2 (i) The nonparametric spectral density estimator $\hat{\boldsymbol{f}}_n(0) = (\hat{\boldsymbol{f}}_{ij}(0))_{i,j=1,\ldots,a}$ of $\boldsymbol{f}(0)$ is consistent.

(ii) It holds that rank $\left(\hat{\boldsymbol{V}}_{\mathrm{GALT},n}\right)$ converges in probability to rank $(\boldsymbol{V})$ as $n \to \infty$.

Remark 4.4 Condition (i) ensures that $\hat{\boldsymbol{V}}_{\mathrm{GALT},n}$ is the consistent estimator of $\boldsymbol{V}$. Condition (ii) with the consistency of $\hat{\boldsymbol{V}}_{\mathrm{GALT},n}$ ensures the convergence of the Moore–Penrose inverse matrix $\hat{\boldsymbol{V}}_{\mathrm{GALT},n}^-$ to $\boldsymbol{V}^-$ (see, e.g., Rakocevic, 1997, Theorem 4.2).

Remark 4.5 As for Condition (i), we can construct the consistent estimator $\hat{\boldsymbol{f}}_n(\lambda) = (\hat{\boldsymbol{f}}_{ij}(\lambda))_{i,j=1,\ldots,a}$ of $\boldsymbol{f}(\lambda)$, for example, by the kernel method, which is defined as $\hat{\boldsymbol{f}}_n(\lambda) = (\hat{\boldsymbol{f}}_{ij}(\lambda))_{i,j=1,\ldots,a}$ and

$$\hat{\boldsymbol{f}}_{ij}(\lambda) := \frac{1}{2\pi} \sum_{\{h\in\mathbb{Z};\, |h|\le\min\{n_i,n_j\}-1\}} \omega\left(\frac{h}{M_n}\right) \hat{\boldsymbol{\Gamma}}_{ij}(h) e^{-ih\lambda}, \quad \lambda \in [-\pi, \pi],$$

where, for $h \in \{0, \ldots, \min\{n_i, n_j\} - 1\}$,

$$\hat{\boldsymbol{\Gamma}}_{ij}(h) := \frac{1}{\min\{n_i, n_j\} - |h|} \sum_{t=1}^{\min\{n_i,n_j\}-|h|} (\boldsymbol{Z}_{i(t+h)} - \boldsymbol{Z}_{i.}(\boldsymbol{Z}_{jt} - \boldsymbol{Z}_{j.})^\top,$$

for $h \in \{-\min\{n_i, n_j\} + 1, \ldots, 0\}$,

$$\hat{\boldsymbol{\Gamma}}_{ij}(h) := \frac{1}{\min\{n_i, n_j\} - |h|} \sum_{t=-h+1}^{\min\{n_i,n_j\}} (\boldsymbol{Z}_{i(t+h)} - \boldsymbol{Z}_{i.})(\boldsymbol{Z}_{jt} - \boldsymbol{Z}_{j.})^\top,$$

M_n is a positive sequence such that $M_n \to \infty$ and $M_n/n \to 0$ as $n \to \infty$, $\omega(x) := \int_{-\infty}^{\infty} W(t)e^{\mathrm{i}xt}\mathrm{d}t$, and the function $W(\cdot)$ satisfies the following conditions: $W(\cdot)$ is a real, bounded, non-negative, even function such that $\int_{-\infty}^{\infty} W(t)\mathrm{d}t = 1$ and $\int_{-\infty}^{\infty} W^2(t)\mathrm{d}t < \infty$ with a bounded derivative.

Remark 4.6 Under the independence of groups and $\boldsymbol{f}_{11}(0) = \cdots = \boldsymbol{f}_{aa}(0)$, $\hat{\boldsymbol{V}}_n$ meets Assumption 4.2 (ii). As an illustration, we set $p = 1$ and $a = 3$. Then,

$$\boldsymbol{V} := \begin{pmatrix} 1-1/a & -1/a & -1/a \\ -1/a & 1-1/a & -1/a \\ -1/a & -1/a & 1-1/a \end{pmatrix} 2\pi f(0)$$

and, for matrices

$$\boldsymbol{P} := \begin{pmatrix} 1 & 0 & 0 \\ 0 & 1 & 0 \\ 1 & 1 & 1 \end{pmatrix} \text{ and } \boldsymbol{B} := \begin{pmatrix} 1-1/a & -1/a & -1/a \\ -1/a & 1-1/a & -1/a \end{pmatrix} 2\pi f(0),$$

it holds that

$$\boldsymbol{P}\boldsymbol{V} = \begin{pmatrix} \boldsymbol{B} \\ 0\ 0\ 0 \end{pmatrix}.$$

Also, the matrix $\boldsymbol{P}\hat{\boldsymbol{V}}_n$ takes the form of

$$\begin{pmatrix} \hat{\boldsymbol{B}}_n \\ 0\ 0\ 0 \end{pmatrix},$$

where $\hat{\boldsymbol{B}}_n$ is some $(a-1)$-by-a matrix. Since $\boldsymbol{B}$ has full rank and the set of all full rank $(a-1)$-by-a matrices is open, $\hat{\boldsymbol{B}}_n$ has full rank for large n. Hence, the condition is satisfied.

Remark 4.7 For $\hat{\boldsymbol{V}}_n$ and $\boldsymbol{V}$, we observe that $\sum_{j=1}^{a} \hat{\boldsymbol{V}}_{ij} = \sum_{j=1}^{a} \boldsymbol{V}_{ij} = \boldsymbol{O}_p$ for any $i \in \{1, \ldots, a\}$. By elementary row operations, we have

$$\boldsymbol{P}\hat{\boldsymbol{V}}_n = \begin{pmatrix} \hat{\boldsymbol{V}}_{11} & \cdots & \hat{\boldsymbol{V}}_{1a} \\ \vdots & \ddots & \vdots \\ \hat{\boldsymbol{V}}_{(a-1)1} & \cdots & \hat{\boldsymbol{V}}_{(a-1)a} \\ \boldsymbol{O}_p & \cdots & \boldsymbol{O}_p \end{pmatrix} \text{ and } \boldsymbol{P}\boldsymbol{V} = \begin{pmatrix} \boldsymbol{V}_{11} & \cdots & \boldsymbol{V}_{1a} \\ \vdots & \ddots & \vdots \\ \boldsymbol{V}_{(a-1)1} & \cdots & \boldsymbol{V}_{(a-1)a} \\ \boldsymbol{O}_p & \cdots & \boldsymbol{O}_p \end{pmatrix},$$

where $\boldsymbol{P} = (\boldsymbol{P}_{i,j})_{i,j=1,\ldots,a}$ with $\boldsymbol{P}_{i,j} = \boldsymbol{I}_a$ for $i = j \in \{1, \ldots, a-1\}$ or $i = a$; otherwise $\boldsymbol{P}_{i,j} = \boldsymbol{O}_p$.

If the matrix

$$\begin{pmatrix} \boldsymbol{V}_{11} & \cdots & \boldsymbol{V}_{1a} \\ \vdots & \ddots & \vdots \\ \boldsymbol{V}_{(a-1)1} & \cdots & \boldsymbol{V}_{(a-1)a} \end{pmatrix}$$

has full rank, Assumption 4.2 (ii) is satisfied since the set of all full rank matrices is open.

Under the independence of the groups and $\boldsymbol{f}_{11}(0) = \cdots = \boldsymbol{f}_{aa}(0) = \boldsymbol{f}(0)$, where $\boldsymbol{f}(0)$ is a full rank p-by-p matrix, we have $\boldsymbol{V} = (\boldsymbol{I}_a - \boldsymbol{J}_a/a) \otimes (2\pi \boldsymbol{f}(0))$. Since $\text{rank}((\boldsymbol{I}_a - \boldsymbol{J}_a/a) \otimes (2\pi \boldsymbol{f}(0)) = \text{rank}(\boldsymbol{I}_a - \boldsymbol{J}_a/a) \times \text{rank}(2\pi \boldsymbol{f}(0)) = (a-1)p$, Assumption 4.2 (ii) is verified.

Remark 4.8 From Remarks 4.6 and 4.7, Assumption 4.2 (ii) can be replaced by the condition that the matrix

$$\begin{pmatrix} \boldsymbol{V}_{11} & \cdots & \boldsymbol{V}_{1a} \\ \vdots & \ddots & \vdots \\ \boldsymbol{V}_{(a-1)1} & \cdots & \boldsymbol{V}_{(a-1)a} \end{pmatrix}$$

has full rank.

Following the above discussion, a test statistic for correlated groups is defined as

$$T_{\text{GALT},n} := n(\boldsymbol{Z}_{1.}^{\top} - \boldsymbol{Z}_{..}^{\top}, \ldots, \boldsymbol{Z}_{a.}^{\top} - \boldsymbol{Z}_{..}^{\top})\hat{\boldsymbol{V}}_{\text{GALT},n}^{-}(\boldsymbol{Z}_{1.}^{\top} - \boldsymbol{Z}_{..}^{\top}, \ldots, \boldsymbol{Z}_{a.}^{\top} - \boldsymbol{Z}_{..}^{\top})^{\top}. \tag{4.7}$$

Note that the test statistic $T_{\text{GALT},n}$ is scale-invariant.

Remark 4.9 Related to Remark 4.1, we can obtain $(q!)^{a-1}$ different p-values by changing the order of the time series when the number of time series from the same groups are more than one ($q \geq 2$). Thus, we propose to consider

$$\boldsymbol{Z}_{it} = (z_{it1.}, z_{it2.}, \ldots, z_{itp.})^{\top},$$

where $z_{itk.} := \sum_{s=1}^{q} z_{itks}/q$ for any $k \in \{1, \ldots, p\}$.

Lemmas 4.1 and 4.2 and Remark 4.4 yield the asymptotic null distribution of $T_{\text{GALT},n}$.

Theorem 4.1 *Suppose Assumptions 4.1 and 4.2 hold. Under H_{41}, $T_{\text{GALT},n}$ converges in distribution to the chi-square distribution with r degrees of freedom as $n \to \infty$, where $r = \text{rank}(\boldsymbol{V})$ and $\boldsymbol{V} = (\boldsymbol{V}_{ij})_{i,j=1\ldots,a}$ are given in Lemma 4.1.*

By Theorem 4.1, for $\varphi \in (0, 1)$, a test with asymptotic size φ can be obtained if we reject H_{41} when $T_{\text{GALT},n} \geq \chi^2_{\hat{r}_n}[1-\varphi]$, where $\chi^2_{\hat{r}_n}[1-\varphi]$ denotes the upper φ-percentiles of the chi-square distribution with $\hat{r}_n$ degrees of freedom, where $\hat{r}_n := \text{rank}\left(\hat{\boldsymbol{V}}_{\text{GALT},n}\right)$.

The theoretical power of the test is stated in the next theorem.

Theorem 4.2 *Suppose Assumptions 4.1 and 4.2 hold. Under the alternative K_{41}, the power of the test based on $T_{\text{GALT},n}$ converges to one, as $n \to \infty$, i.e., the test is consistent.*

Proof (Theorem 4.2) Note that $\boldsymbol{V}$ is non-negative definite since $\boldsymbol{V}$ is the asymptotic variance (see Lemma 4.1) and the Moore–Penrose inverse of any non-negative definite matrix is non-negative definite (Wu, 1980, Theorem 1). Thus, we observe, under the alternative K_{41}, that

$$\begin{aligned}
&\mathrm{P}\left(T_{\mathrm{GALT},n} \geq \chi^2_{\hat{r}_n}[1-\varphi]\right) \\
&= \mathrm{P}\left(\left\{\begin{pmatrix}\boldsymbol{\alpha}_1\\ \vdots \\ \boldsymbol{\alpha}_a\end{pmatrix} + \begin{pmatrix}\boldsymbol{e}_{1.}-\boldsymbol{e}_{..}\\ \vdots \\ \boldsymbol{e}_{a.}-\boldsymbol{e}_{..}\end{pmatrix}\right\}^\top \hat{\boldsymbol{V}}^-_{\mathrm{GALT},n}\left\{\begin{pmatrix}\boldsymbol{\alpha}_1\\ \vdots \\ \boldsymbol{\alpha}_a\end{pmatrix} + \begin{pmatrix}\boldsymbol{e}_{1.}-\boldsymbol{e}_{..}\\ \vdots \\ \boldsymbol{e}_{a.}-\boldsymbol{e}_{..}\end{pmatrix}\right\} \geq \frac{\chi^2_{\hat{r}_n}[1-\varphi]}{n}\right) \\
&\to \mathrm{P}\left((\boldsymbol{\alpha}_1^\top,\ldots,\boldsymbol{\alpha}_a^\top)\boldsymbol{V}^-(\boldsymbol{\alpha}_1^\top,\ldots,\boldsymbol{\alpha}_a^\top)^\top \geq 0\right) = 1
\end{aligned}$$

as $n \to \infty$.

Next, we consider the local alternative hypothesis to provide a benchmark for the comparison of the test based on $T_{\mathrm{GALT},n}$ with other consistent tests. The local alternative hypothesis is defined, for the perturbations $\boldsymbol{h}_1, \ldots, \boldsymbol{h}_a$ satisfying $\sum_{i=1}^{a} \boldsymbol{h}_i = \boldsymbol{0}$, as

$$K_{41}^{(n)} : \boldsymbol{\alpha}_i = \frac{\boldsymbol{h}_i}{\sqrt{n}} \quad (i = 1, \ldots, a).$$

Then, the nontrivial power under the local alternative is elucidated in the following theorem.

Theorem 4.3 *Suppose Assumptions 4.1 and 4.2 hold. Under the local alternatives $K_{41}^{(n)}$, $T_{\mathrm{GALT},n}$ converges in distribution to the noncentral chi-square distribution with r degrees of freedom and the noncentrality parameter $\delta = (\boldsymbol{h}_1^\top, \ldots, \boldsymbol{h}_a^\top)\boldsymbol{V}^-(\boldsymbol{h}_1^\top, \ldots, \boldsymbol{h}_a^\top)^\top$, as $n \to \infty$.*

Proof (Theorem 4.3)
Under the local alternative, $K_{41}^{(n)}$, a simple algebra gives

$$T_{\mathrm{GALT},n} = \left\{\begin{pmatrix}\boldsymbol{h}_1\\ \vdots \\ \boldsymbol{h}_a\end{pmatrix} + \sqrt{n}\begin{pmatrix}\boldsymbol{e}_{1.}-\boldsymbol{e}_{..}\\ \vdots \\ \boldsymbol{e}_{a.}-\boldsymbol{e}_{..}\end{pmatrix}\right\} \hat{\boldsymbol{V}}^-_{\mathrm{GALT},n}\left\{\begin{pmatrix}\boldsymbol{h}_1\\ \vdots \\ \boldsymbol{h}_a\end{pmatrix} + \sqrt{n}\begin{pmatrix}\boldsymbol{e}_{1.}-\boldsymbol{e}_{..}\\ \vdots \\ \boldsymbol{e}_{a.}-\boldsymbol{e}_{..}\end{pmatrix}\right\}^\top,$$

which, combined with Lemmas 4.1 and 4.2, shows that $T_{\mathrm{GALT},n}$ converges in distribution to the noncentral chi-square distribution with r degrees of freedom and noncentrality parameter $\delta = (\boldsymbol{h}_1^\top, \ldots, \boldsymbol{h}_a^\top)\boldsymbol{V}^-(\boldsymbol{h}_1^\top, \ldots, \boldsymbol{h}_a^\top)^\top$.

Theorem 4.3 shows that the nontrivial power of the test under the local alternatives can be expressed as

$$1 - \Psi_{r,\delta}(\chi^2_r[1-\varphi]),$$

where $\Psi_{r,\delta}$ is the cumulative distribution function of the noncentral chi-square distribution with r degrees of freedom and the noncentrality parameter δ.

Remark 4.10 A parametric spectral density matrix $\boldsymbol{f}_{\boldsymbol{\theta}_0}(\lambda)$ can be considered instead of the nonparametric spectral density matrix $\boldsymbol{f}(\lambda)$. The unknown parameter $\boldsymbol{\theta}_0 \in \mathbb{R}^d$ can be estimated by the minimum discrepancy estimator. For the family of parametric spectral density $\{\boldsymbol{f}_{\boldsymbol{\theta}}(\lambda); \boldsymbol{\theta} \in \boldsymbol{\Theta} \subset \mathbb{R}^d\}$, the estimator is defined as

$$\hat{\boldsymbol{\theta}}_n := \arg\min_{\boldsymbol{\theta} \in \boldsymbol{\Theta}} \int_{-\pi}^{\pi} \mathfrak{K}\{\boldsymbol{\theta}, \hat{\boldsymbol{f}}_n(\lambda), \lambda\} \mathrm{d}\lambda,$$

where $\mathfrak{K}$ is an appropriate function. If we choose $\mathfrak{K}$ as

$$\mathfrak{K}\{\boldsymbol{\theta}, \hat{\boldsymbol{f}}_n(\lambda), \lambda\} = -\log\det(\hat{\boldsymbol{f}}_n(\lambda)\boldsymbol{f}_{\boldsymbol{\theta}}^{-1}(\lambda)) + \mathrm{tr}(\hat{\boldsymbol{f}}_n(\lambda)\boldsymbol{f}_{\boldsymbol{\theta}}^{-1}(\lambda)) - q,$$

this estimator reduces to the Whittle likelihood estimator. By Taniguchi and Kakizawa (2000, Theorem 6.2.3) with Robinson (1991, Theorem 2.1), we derive the consistency of $\hat{\boldsymbol{\theta}}_n$ under appropriate conditions. If $\boldsymbol{f}_{\boldsymbol{\theta}}$ is continuous with respect to $\boldsymbol{\theta}$, we can show that $f_{\hat{\boldsymbol{\theta}}_n}(\lambda)$ converges in probability to $f_{\boldsymbol{\theta}_0}(\lambda)$. Therefore, the testing theory stated in this section remains valid even if we replace $\hat{\boldsymbol{f}}_n(0)$ in $\hat{\boldsymbol{V}}^{-}_{\mathrm{GALT},n}$ with $f_{\hat{\boldsymbol{\theta}}_n}(0)$.

4.2 Test for the Existence of Random Effects

In this section, we consider a test for the existence of random effects for one-way random effect models with time-dependent errors and correlated groups.

The random effect model is the same as the fixed effect model (4.1) but $\boldsymbol{\alpha}_i$ is a random effect of the i-th group. For simplicity, we assume $(\boldsymbol{\alpha}_1^\top, \ldots, \boldsymbol{\alpha}_a^\top)^\top$ follows the ap-dimensional centered normal distribution with variance ${}^{\alpha}\boldsymbol{\Sigma} = ({}^{\alpha}\boldsymbol{\Sigma}_{ij})_{i,j=1,\ldots,a}$ and the $\{\boldsymbol{\alpha}_j\}$'s are independent of $\{\boldsymbol{e}_{it}; t = 1, ..., n_i\}$. Note that the spectral density of $\boldsymbol{Z}_{it}$ does not exist for the random effect model.

The null hypothesis H_{42} and the alternative K_{42} for the existence of random effects can be formulated as follows:

$$H_{42} : {}^{\alpha}\boldsymbol{\Sigma} = \boldsymbol{O}_{ap} \quad \text{vs} \quad K_{42} : {}^{\alpha}\boldsymbol{\Sigma} \neq \boldsymbol{O}_{ap},$$

where $\boldsymbol{O}_{ap}$ is an ap-by-ap zero matrix.

Remark 4.11 The Gaussian assumption on the random effects is not essential. For a centered non-Gaussian random vector $(\boldsymbol{\alpha}_1^\top, \ldots, \boldsymbol{\alpha}_a^\top)^\top$, the null and alternative hypothesis can be defined as

$$H_{43} : \mathrm{P}(\boldsymbol{\alpha}_1 = \cdots = \boldsymbol{\alpha}_a) = 1 \quad \text{vs}$$
$$K_{43} : \mathrm{P}(\boldsymbol{\alpha}_1 = \cdots = \boldsymbol{\alpha}_a) = 0.$$

If $\{\boldsymbol{\alpha}_i\}$ is an i.i.d. continuous random variable with respect to i, K_{43} is always satisfied. On the other hand, K_{43} is not fulfilled when $\{\boldsymbol{\alpha}_i\}$ is an i.i.d. discrete random variable with respect to i.

The test statistic $T_{\mathrm{GALT},n}$, defined in (4.7), is still available for the random effect model. The asymptotic null distribution is exactly the same as that for the fixed effect model.

Theorem 4.4 *Suppose Assumptions 4.1 and 4.2 hold. Under the null H_{42}, $T_{\mathrm{GALT},n}$ converges in distribution to the chi-square distribution with r degrees of freedom as $n \to \infty$.*

Proof (Theorem 4.4) The random effect model under the null H_{42} is equal to the fixed effect model under the null H_{41}. Hence, the proof is the same as Theorem 4.1.

Thus, a test with asymptotic size α can be derived if we reject H_{42} whenever $T_{\mathrm{GALT},n} \geq \chi^2_{\hat{r}_n}[1-\varphi]$. The consistency of the test is given in the following theorem.

Theorem 4.5 *Suppose Assumptions 4.1 and 4.2 hold. The test based on $T_{\mathrm{GALT},n}$ is consistent. More precisely, under the alternative K_{42}, it holds that $\mathrm{P}(T_{\mathrm{GALT},n} \geq \chi^2_{\hat{r}_n}[1-\varphi]) \to 1$, as $n \to \infty$.*

Proof (Theorem 4.5) We note that the Moore–Penrose inverse of any non-negative definite matrix is non-negative definite (Wu, 1980, Theorem 1), $\boldsymbol{V}$ is the asymptotic variance, and, thus, $\boldsymbol{V}$ is non-negative definite. Then, we know that, under the alternative K_{42},

$$\begin{aligned}
&\mathrm{P}\left(\frac{T_{\mathrm{GALT},n}}{n} \geq \frac{\chi^2_{\hat{r}_n}[1-\varphi]}{n}\right)\\
=&\mathrm{P}\Big((\boldsymbol{\alpha}_1^\top - \boldsymbol{\alpha}_\cdot^\top + \boldsymbol{e}_{1\cdot}^\top - \boldsymbol{e}_{\cdot\cdot}^\top, \ldots, \boldsymbol{\alpha}_a^\top - \boldsymbol{\alpha}_\cdot^\top + \boldsymbol{e}_{a\cdot}^\top - \boldsymbol{e}_{\cdot\cdot}^\top)\hat{\boldsymbol{V}}_n^-\\
&\quad \times \left(\boldsymbol{\alpha}_1^\top - \boldsymbol{\alpha}_\cdot^\top + \boldsymbol{e}_{1\cdot}^\top - \boldsymbol{e}_{\cdot\cdot}^\top, \ldots, \boldsymbol{\alpha}_a^\top - \boldsymbol{\alpha}_\cdot^\top + \boldsymbol{e}_{a\cdot}^\top - \boldsymbol{e}_{\cdot\cdot}^\top\right)^\top \geq \frac{\chi^2_{\hat{r}_n}[1-\varphi]}{n}\Big)\\
=&\mathrm{P}\Big(\left(\boldsymbol{\alpha}_1^\top - \boldsymbol{\alpha}_\cdot^\top, \ldots, \boldsymbol{\alpha}_a^\top - \boldsymbol{\alpha}_\cdot^\top\right)\boldsymbol{V}^-\left(\boldsymbol{\alpha}_1^\top - \boldsymbol{\alpha}_\cdot^\top, \ldots, \boldsymbol{\alpha}_a^\top - \boldsymbol{\alpha}_\cdot^\top\right)^\top \geq 0\Big) + o(1)\\
\to&\mathrm{P}\Big(N\left(\boldsymbol{0}, {}^{\alpha}\tilde{\boldsymbol{\Sigma}}\right)\boldsymbol{V}^- N\left(\boldsymbol{0}, {}^{\alpha}\tilde{\boldsymbol{\Sigma}}\right)^\top \geq 0\Big) = 1
\end{aligned}$$

as $n \to \infty$, where ${}^{\alpha}\tilde{\boldsymbol{\Sigma}} = ({}^{\alpha}\tilde{\boldsymbol{\Sigma}}_{ij})_{i,j=1\ldots,a}$ is given by

$${}^{\alpha}\tilde{\boldsymbol{\Sigma}}_{ij} = {}^{\alpha}\boldsymbol{\Sigma}_{ij} - \frac{1}{a}\sum_{s=1}^{a}({}^{\alpha}\boldsymbol{\Sigma}_{sj} + {}^{\alpha}\boldsymbol{\Sigma}_{is}) + \frac{1}{a^2}\sum_{s,k=1}^{a}{}^{\alpha}\boldsymbol{\Sigma}_{sk}.$$

Before ending this chapter, we derive the nontrivial power of the test based on $T_{\mathrm{GALT},n}$ under the local alternative hypothesis. Let $\boldsymbol{H} = (\boldsymbol{H}_{ij})_{i,j=1\ldots,a}$ be an ap-by-ap symmetric, positive definite matrix, and the local alternatives $K_{42}^{(n)}$ be defined as

$$K_{42}^{(n)} : {}^{\boldsymbol{\alpha}}\boldsymbol{\Sigma} = \frac{\boldsymbol{H}}{n}.$$

The nontrivial power of the test is elucidated in the following theorem.

Theorem 4.6 *Suppose Assumptions 4.1 and 4.2 hold. Under the alternatives $K_{42}^{(n)}$, we have*

$$\lim_{n\to\infty} \mathrm{P}(T_{\mathrm{GALT},n} \geq \chi^2_{\hat{r}_n}[1-\varphi]) = \mathrm{P}\left(\boldsymbol{G}^\top \boldsymbol{V}^- \boldsymbol{G} \geq \chi^2_r[1-\varphi]\right),$$

where $\boldsymbol{G}$ follows an ap-dimensional centered normal distribution with variance $\tilde{\boldsymbol{H}} + \boldsymbol{V}$ and $\tilde{\boldsymbol{H}} = (\tilde{\boldsymbol{H}}_{ij})_{i,j=1\ldots,a}$ is defined as

$$\tilde{\boldsymbol{H}}_{ij} := \boldsymbol{H}_{ij} - \frac{1}{a}\sum_{s=1}^{a}(\boldsymbol{H}_{sj} + \boldsymbol{H}_{is}) + \frac{1}{a^2}\sum_{s,k=1}^{a} \boldsymbol{H}_{sk}.$$

Proof (Theorem 4.6) We observe that, under the local alternative $K_{42}^{(n)}$,

$$\begin{aligned}
&\mathrm{P}_n(T_{\mathrm{GALT},n} \geq \chi^2_{\hat{r}_n}[1-\varphi]) \\
=&\mathrm{P}_n\Big(\big(\sqrt{n}(\boldsymbol{\alpha}_1^\top - \boldsymbol{\alpha}_{\cdot}^\top + \boldsymbol{e}_{1\cdot}^\top - \boldsymbol{e}_{\cdot\cdot}^\top), \ldots, \sqrt{n}(\boldsymbol{\alpha}_a^\top - \boldsymbol{\alpha}_{\cdot}^\top + \boldsymbol{e}_{a\cdot}^\top - \boldsymbol{e}_{\cdot\cdot}^\top)\big) \hat{\boldsymbol{V}}_n^- \\
&\times \big(\sqrt{n}(\boldsymbol{\alpha}_1^\top - \boldsymbol{\alpha}_{\cdot}^\top + \boldsymbol{e}_{1\cdot}^\top - \boldsymbol{e}_{\cdot\cdot}^\top), \ldots, \sqrt{n}(\boldsymbol{\alpha}_a^\top - \boldsymbol{\alpha}_{\cdot}^\top + \boldsymbol{e}_{a\cdot}^\top - \boldsymbol{e}_{\cdot\cdot})^\top\big)^\top \\
&\geq \chi^2_{\hat{r}_n}[1-\varphi]\Big) \\
\to&\mathrm{P}\left(N(\boldsymbol{0}, \tilde{\boldsymbol{H}} + \boldsymbol{V})\boldsymbol{V}^- N(\boldsymbol{0}, \tilde{\boldsymbol{H}} + \boldsymbol{V})^\top \geq \chi^2_r[1-\varphi]\right)
\end{aligned}$$

as $n \to \infty$, where P_n is a sequence of the probability distributions for $K_{42}^{(n)}$.

Remark 4.12 We can generalize the random effects $(\boldsymbol{\alpha}_1^\top, \ldots, \boldsymbol{\alpha}_a^\top)^\top$ to an ap-dimensional random vector, and show theoretical results corresponding to Theorems 4.4–4.6 for the hypothesis given in Remark 4.11.

References

Brillinger, D. R. (1981). *Time series: Data analysis and theory*. San Francisco: Holden-Day.

Goto, Y., Arakaki, K., Liu, Y., & Taniguchi, M. (2023). Homogeneity tests for one-way models with dependent errors. *TEST, 32*, 163–183.

Hosoya, Y., & Taniguchi, M. (1982). A central limit theorem for stationary processes and the parameter estimation of linear processes. *Ann. Statist.* 132–153.

Rakocevic, V. (1997). On continuity of the Moore-Penrose and Drazin inverses. *Mat. Vesn., 49*, 163–172.

Rao, C. R., & Mitra, S. K. (1971). *Generalized inverse of matrices and its applications*. John Wiley and Sons, Inc.

Robinson, P. M. (1991). Automatic frequency domain inference on semiparametric and nonparametric models. *Econometrica, 59*, 1329–1363.

Taniguchi, M., & Kakizawa, Y. (2000). *Asymptotic theory of statistical inference for time series*. Springer Science & Business Media.

Wu, C.-F. (1980). On some ordering properties of the generalized inverses of nonnegative definite matrices. *Linear Algebra Its Appl., 32*, 49–60.

Chapter 5
Two-Way Random Effect Model for Correlated Cells

This chapter extends the one-way models with time-dependent errors and correlated groups dealt with in Chapter 4 to the two-way models with time-dependent errors and correlated groups. We propose a test for the existence of random effects in Section 5.1 and a test for the existence of random interactions in Section 5.2. We show that the tests have asymptotically size φ and are consistent. The nontrivial power under the local alternative is elucidated. These tests can be applied to the fixed and mixed effect models. This chapter is mainly based on Goto et al. (2023).

5.1 Test for the Existence of Random Effects

Two-way random effect models with time-dependent errors and correlated groups are defined as

$$\boldsymbol{Z}_{ijt} = \boldsymbol{\mu} + \boldsymbol{\alpha}_i + \boldsymbol{\beta}_j + \boldsymbol{e}_{ijt}, \qquad i = 1, \ldots, a;\ j = 1, \ldots, b;\ t = 1, \ldots, n_{ij}, \quad (5.1)$$

where $\boldsymbol{Z}_{ijt} := (z_{ijt1}, \ldots, z_{ijtp})^\top$ is a t-th p-dimensional observation in the (i, j)-th cell, $\boldsymbol{\mu} := (\mu_1, \ldots, \mu_p)^\top$ is a p-dimensional grand mean, $\boldsymbol{\alpha}_i := (\alpha_{i1}, \ldots, \alpha_{ip})^\top$ and $\boldsymbol{\beta}_j := (\beta_{j1}, \ldots, \beta_{jp})^\top$ are p-dimensional random effects of the i-th level of factor A and the j-th level of factor B, respectively, and $\boldsymbol{e}_{ijt} := (e_{ijt1}, \ldots, e_{ijtp})^\top$ is a centered stationary sequence.

We suppose that

(i) the observed stretch $\{\boldsymbol{Z}_{ijt};\ i = 1, \ldots, a;\ j = 1, \ldots, b; t = 1, \ldots, n_{ij}\}$ is available,
(ii) any two of $\{\boldsymbol{\alpha}_i\}$, $\{\boldsymbol{\beta}_j\}$, and $\{\boldsymbol{e}_{ijt}\}$ are independent,
(iii) the centered stationary time series

$$(\boldsymbol{e}_{11t}^\top, \boldsymbol{e}_{21t}^\top, \ldots, \boldsymbol{e}_{a1t}^\top, \boldsymbol{e}_{12t}^\top, \ldots, \boldsymbol{e}_{a2t}^\top, \ldots, \boldsymbol{e}_{1bt}^\top, \ldots, \boldsymbol{e}_{abt}^\top)^\top$$

Y. Goto et al., *ANOVA with Dependent Errors*, JSS Research Series in Statistics,
https://doi.org/10.1007/978-981-99-4172-8_5

has an abp-by-abp spectral density matrix $\boldsymbol{f}(\lambda) := \{\boldsymbol{f}_{j_1j_2}(\lambda)\}_{j_1,j_2=1,\dots,b}$ with $\boldsymbol{f}_{j_1j_2}(\lambda) := \{\boldsymbol{f}^{i_1i_2}_{j_1j_2}(\lambda)\}_{i_1,i_2=1,\dots,a}$ and

$$\boldsymbol{f}^{i_1i_2}_{j_1j_2}(\lambda) := \sum_{k=-\infty}^{\infty} \mathrm{E}(\boldsymbol{e}_{i_1j_1t+k}\boldsymbol{e}^{\top}_{i_2j_2t})e^{-ik\lambda}/(2\pi),$$

and

(iv) there exist some constants $\rho_{ij} \in (0,1)$ such that $n_{ij} = \rho_{ij}n$ with $n = \sum_{i=1}^{a}\sum_{j=1}^{b} n_{ij}$ for any $(i,j) \in \{1,\dots,a\} \times \{1,\dots,b\}$.

Remark 5.1 Condition (iii) allows to deal with correlated cells and (iv) covers the case when the length of the time series varies from cell to cell.

Remark 5.2 In the same way as Remark 4.10, we can consider the parametric spectral density matrix.

Moreover, we assume that, for simplicity, $(\boldsymbol{\alpha}_1^{\top},\dots,\boldsymbol{\alpha}_a^{\top})^{\top}$ and $(\boldsymbol{\beta}_1^{\top},\dots,\boldsymbol{\beta}_b^{\top})^{\top}$ follow ap- and bp-dimensional centered normal distributions with variances ${}^{\alpha}\boldsymbol{\Sigma} := ({}^{\alpha}\boldsymbol{\Sigma}_{i_1i_2})_{i_1,i_2=1,\dots,a}$ and ${}^{\beta}\boldsymbol{\Sigma} := ({}^{\beta}\boldsymbol{\Sigma}_{j_1j_2})_{j_1,j_2=1,\dots,b}$, respectively, where ${}^{\alpha}\boldsymbol{\Sigma}_{i_1i_2} := \mathrm{E}(\boldsymbol{\alpha}_{i_1}\boldsymbol{\alpha}^{\top}_{i_2})$ and ${}^{\beta}\boldsymbol{\Sigma}_{j_1j_2} := \mathrm{E}(\boldsymbol{\beta}_{j_1}\boldsymbol{\beta}^{\top}_{j_2})$.

The tests for the random effects $\boldsymbol{\alpha}_i$ and $\boldsymbol{\beta}_j$ can be formulated as

$$H_{51} : {}^{\alpha}\boldsymbol{\Sigma} = \boldsymbol{O}_{ap} \quad \text{vs} \quad K_{51} : {}^{\alpha}\boldsymbol{\Sigma} \neq \boldsymbol{O}_{ap} \tag{5.2}$$

and

$$H_{52} : {}^{\beta}\boldsymbol{\Sigma} = \boldsymbol{O}_{bp} \text{ vs } K_{52} : {}^{\beta}\boldsymbol{\Sigma} \neq \boldsymbol{O}_{bp}, \tag{5.3}$$

respectively. We focus on the hypothesis (5.2) since we can construct a test for the hypothesis (5.3) in the same manner (see Remark 5.6).

In the same spirit of Chapter 4, we consider the test statistic standardized not only within cells but also between cells to cope with correlated cells. We make the following assumption.

Assumption 5.1 For all $\ell \in \mathbb{N}, i_1,\dots,i_\ell \in \{1,\dots,a\}, j_1,\dots,j_\ell \in \{1,\dots,b\}$, and $d_1,\dots,d_\ell \in \{1,\dots,p\}$, it holds that

$$\sum_{s_2,\dots,s_\ell=-\infty}^{\infty} \left\{\left(1+\sum_{k=2}^{\ell}|s_k|\right)\left|\varkappa^{i_1\cdots i_\ell}_{j_1\cdots j_\ell}(s_2,\dots,s_\ell;d_1,\dots,d_\ell)\right|\right\} < \infty,$$

where $\varkappa^{i_1\cdots i_\ell}_{j_1\cdots j_\ell}(s_2,\dots,s_\ell;d_1,\dots,d_\ell) := \mathrm{Cum}\{e_{i_1j_10d_1}, e_{i_2j_2s_2d_2},\dots,e_{i_\ell j_\ell s_\ell d_\ell}\}$ and, for a random variable $\{X_t\}$, $\mathrm{Cum}(X_1,\dots,X_\ell)$ denotes the cumulant of order ℓ of $(X_1,\dots,X_\ell)$ (see Brillinger, 1981, p. 19).

The next lemma gives an intuition on how to construct the test statistic.

Lemma 5.1 *Suppose Assumption 5.1 holds. Under* H_{51},

$$\sqrt{n}\left(\boldsymbol{Z}_{1..}^{\top}-\boldsymbol{Z}_{...}^{\top}\ \ldots\ \boldsymbol{Z}_{a..}^{\top}-\boldsymbol{Z}_{...}^{\top}\right)^{\top}$$

converges in distribution to the ap-dimensional centered normal distribution with variance $\boldsymbol{V}_{\boldsymbol{\alpha}}$ *as* $\min_{\substack{i=1,\ldots,a\\ j=1,\ldots,b}} n_{ij}\to\infty$, *where* $\boldsymbol{V}_{\boldsymbol{\alpha}}:=({}^{\boldsymbol{\alpha}}\boldsymbol{V}_{i_1i_2})_{i_1,i_2=1,\ldots,a}$,

$$^{\boldsymbol{\alpha}}\boldsymbol{V}_{i_1i_2}:=\boldsymbol{\zeta}_{..}^{i_1i_2}-\boldsymbol{\zeta}_{..}^{i_1\cdot}-\boldsymbol{\zeta}_{..}^{\cdot i_2}+\boldsymbol{\zeta}_{..}^{..},\quad \boldsymbol{\zeta}_{j_1j_2}^{i_1i_2}:=2\pi\frac{\min\{\rho_{i_1j_1},\rho_{i_2j_2}\}}{\rho_{i_1j_1}\rho_{i_2j_2}}\boldsymbol{f}_{j_1j_2}^{i_1i_2}(0),$$

and a subscript dot denotes taking the average with respect to the corresponding element, for example, $\boldsymbol{\zeta}_{..}^{i_1i_2}:=\sum_{j_1,j_2=1}^{b}\boldsymbol{\zeta}_{j_1j_2}^{i_1i_2}/b^2$.

***Proof (Lemma** 5.1)* For any $i\in\{1,\ldots,a\}$, under H_{51} we have $\sqrt{n}\mathrm{E}(\boldsymbol{Z}_{i..}-\boldsymbol{Z}_{...})=\boldsymbol{0}$. Under H_{51}, we observe that, for any $i_1,i_2\in\{1,\ldots,a\}$,

$$n\mathrm{Cov}(\boldsymbol{Z}_{i_1..}-\boldsymbol{Z}_{...},\boldsymbol{Z}_{i_2..}-\boldsymbol{Z}_{...})=n\mathrm{Cov}(\boldsymbol{e}_{i_1..}-\boldsymbol{e}_{...},\boldsymbol{e}_{i_2..}-\boldsymbol{e}_{...})={}^{n,\boldsymbol{\alpha}}\boldsymbol{V}_{i_1i_2},$$

where

$$\begin{aligned}
{}^{n,\boldsymbol{\alpha}}\boldsymbol{V}_{i_1i_2}:=&\frac{1}{nb^2}\sum_{j_1,j_2=1}^{b}\frac{1}{\rho_{i_1j_1}\rho_{i_2j_2}}\sum_{t_1=1}^{n_{i_1j_1}}\sum_{t_2=1}^{n_{i_2j_2}}\boldsymbol{\Gamma}_{j_1j_2}^{i_1i_2}(t_1-t_2)\\
&-\frac{1}{nab^2}\sum_{s=1}^{a}\sum_{j_1,j_2=1}^{b}\sum_{t_1=1}^{n_{sj_1}}\sum_{t_2=1}^{n_{i_2j_2}}\frac{1}{\rho_{sj_1}\rho_{i_2j_2}}\boldsymbol{\Gamma}_{j_1j_2}^{si_2}(t_1-t_2)\\
&-\frac{1}{nab^2}\sum_{s=1}^{a}\sum_{j_1,j_2=1}^{b}\sum_{t_1=1}^{n_{i_1j_1}}\sum_{t_2=1}^{n_{sj_2}}\frac{1}{\rho_{i_1j_1}\rho_{sj_2}}\boldsymbol{\Gamma}_{j_1j_2}^{i_1s}(t_1-t_2)\\
&+\frac{1}{na^2b^2}\sum_{s_1,s_2=1}^{a}\sum_{j_1,j_2=1}^{b}\frac{1}{\rho_{s_1j_1}\rho_{s_2j_2}}\sum_{t_1=1}^{n_{s_1j_1}}\sum_{t_2=1}^{n_{s_2j_2}}\boldsymbol{\Gamma}_{j_1j_2}^{s_1s_2}(t_1-t_2)
\end{aligned}$$

with $\boldsymbol{\Gamma}_{j_1j_2}^{i_1,i_2}(h):=\mathrm{E}(\boldsymbol{e}_{i_1j_1t+h}\boldsymbol{e}_{i_2j_2t}^{\top})$. From Assumption 5.1, ${}^{n,\boldsymbol{\alpha}}\boldsymbol{V}_{i_1i_2}$ converges to ${}^{\boldsymbol{\alpha}}\boldsymbol{V}_{i_1i_2}$ as $\min_{\substack{i=1,\ldots,a\\ j=1,\ldots,b}} n_{ij}\to\infty$. For any $\ell\geq 3$, $(i_1,\ldots,i_\ell)\in\{1,\ldots,a\}^\ell$, $(j_1,\ldots,j_\ell)\in\{1,\ldots,b\}^\ell$, and $(d_1,\ldots,d_\ell)\in\{1,\ldots,p\}^\ell$, Assumption 5.1 yields

$$\begin{aligned}
&n^{\ell/2}\mathrm{Cum}\{e_{i_1j_1.d_1},\ldots,e_{i_\ell j_\ell.d_\ell}\}\\
&=\frac{n^{-\ell/2}}{\rho_{i_1j_1}\cdots\rho_{i_\ell j_\ell}}\sum_{t_1=1}^{n_{i_1j_1}}\cdots\sum_{t_\ell=1}^{n_{i_\ell j_\ell}}|\mathrm{Cum}\{e_{i_1j_1t_1d_1},\ldots,e_{i_\ell j_\ell t_\ell d_\ell}\}|\\
&=\frac{n^{-\ell/2}}{\rho_{i_1j_1}\cdots\rho_{i_\ell j_\ell}}\sum_{t_1=1}^{n_{i_1j_1}}\sum_{s_2=1-t_1}^{n_{i_2j_2}-t_1}\cdots\sum_{s_\ell=1-t_1}^{n_{i_\ell j_\ell}-t_1}|\kappa_{i_1\cdots i_\ell}^{j_1\cdots j_\ell}(s_2,\ldots,s_\ell;d_1,\ldots,d_\ell)|
\end{aligned}$$

$$\leq \frac{n^{-\ell/2}}{\rho_{i_1 j_1} \cdots \rho_{i_\ell j_\ell}} \sum_{t_1=1}^{n_{i_1 j_1}} \sum_{s_2=-\infty}^{\infty} \cdots \sum_{s_\ell=-\infty}^{\infty} |\kappa_{i_1 \cdots i_\ell}^{j_1 \cdots j_\ell}(s_2, \ldots, s_\ell; d_1, \ldots, d_\ell)|$$
$$= \frac{n^{-\ell/2+1}}{\rho_{i_2 j_2} \cdots \rho_{i_\ell j_\ell}} \sum_{s_2=-\infty}^{\infty} \cdots \sum_{s_\ell=-\infty}^{\infty} |\kappa_{i_1 \cdots i_\ell}^{j_1 \cdots j_\ell}(s_2, \ldots, s_\ell; d_1, \ldots, d_\ell)|$$
$$= O(n^{-\ell/2+1}), \tag{5.4}$$

where $e_{ij.d} := \sum_{t=1}^{n_{ij}} e_{ijtd}/n_{ij}$ for any $(i, j, d) \in \{1, \ldots, a\} \times \{1, \ldots, b\} \times \{1, \ldots, p\}$. □

Motivated by Lemma 5.1, the test statistic for (5.2) is defined as

$$T_{\boldsymbol{\alpha},\mathrm{GSXT},n} := \begin{pmatrix} \boldsymbol{Z}_{1..} - \boldsymbol{Z}_{...} \\ \vdots \\ \boldsymbol{Z}_{a..} - \boldsymbol{Z}_{...} \end{pmatrix} \hat{\boldsymbol{V}}_{n,\boldsymbol{\alpha}}^{-} \begin{pmatrix} \boldsymbol{Z}_{1..} - \boldsymbol{Z}_{...} \\ \vdots \\ \boldsymbol{Z}_{a..} - \boldsymbol{Z}_{...} \end{pmatrix}^{\top}, \tag{5.5}$$

where $\hat{\boldsymbol{V}}_{n,\boldsymbol{\alpha}}^{-}$ is the Moore–Penrose inverse matrix of $\hat{\boldsymbol{V}}_{n,\boldsymbol{\alpha}} := ({}^{\boldsymbol{\alpha}}\hat{\boldsymbol{V}}_{i_1 i_2})_{i_1,i_2=1,\ldots,a}$,

$${}^{\boldsymbol{\alpha}}\hat{\boldsymbol{V}}_{i_1 i_2} := \hat{\boldsymbol{\zeta}}_{..}^{i_1 i_2} - \hat{\boldsymbol{\zeta}}_{..}^{i_1 \cdot} - \hat{\boldsymbol{\zeta}}_{..}^{\cdot i_2} + \hat{\boldsymbol{\zeta}}_{..}^{..}, \quad \hat{\boldsymbol{\zeta}}_{j_1 j_2}^{i_1 i_2} := \frac{2\pi \min\{\rho_{i_1 j_1}, \rho_{i_2 j_2}\}}{\rho_{i_1 j_1} \rho_{i_2 j_2}} \hat{\boldsymbol{f}}_{j_1 j_2}^{i_1 i_2}(0),$$

and $\hat{\boldsymbol{f}}_n(\lambda) := \left(\hat{\boldsymbol{f}}_{j_1 j_2}(\lambda)\right)_{j_1,j_2=1,\ldots,b}$ with its entries $\hat{\boldsymbol{f}}_{j_1 j_2}(\lambda) := \left(\hat{f}_{j_1 j_2}^{i_1 i_2}(\lambda)\right)_{i_1,i_2=1,\ldots,a}$ is a nonparametric spectral density matrix estimator of $\boldsymbol{f}(\lambda)$.

Since $\sum_{i_1=1}^{a} {}^{\boldsymbol{\alpha}}\hat{\boldsymbol{V}}_{i_1 i_2} = \sum_{i_1=1}^{a} {}^{\boldsymbol{\alpha}}\boldsymbol{V}_{i_1 i_2} = \boldsymbol{O}_p$, $\hat{\boldsymbol{V}}_{n,\boldsymbol{\alpha}}^{-}$ and $\boldsymbol{V}_{\boldsymbol{\alpha}}$ are always singular and, thus, adopting the Moore–Penrose inverse instead of the usual inverse is essential.

In order to obtain the basic properties of the test based on $T_{\boldsymbol{\alpha},\mathrm{GSXT},n}$, we make the following assumptions.

Assumption 5.2 (i) The nonparametric spectral density estimator $\hat{\boldsymbol{f}}_n(0)$ of $\boldsymbol{f}(0)$ is consistent.

(ii) It holds that

$$\mathrm{rank}\left(\hat{\boldsymbol{V}}_{n,\boldsymbol{\alpha}}\right) \to \mathrm{rank}\left(\boldsymbol{V}_{\boldsymbol{\alpha}}\right) \quad \text{in probability, as} \min_{\substack{i=1,\ldots,a \\ j=1,\ldots,b}} n_{ij} \to \infty.$$

Remark 5.3 As for Assumption (i), we can construct the consistent estimator $\hat{\boldsymbol{f}}_n(0)$ of $\boldsymbol{f}(0)$ by the kernel method (see also Remark 4.5). Assumption (ii) ensures the convergence of the Moore–Penrose inverse matrix (see, e.g., Rakocevic, 1997, Theorem 4.2).

By applying Lemma 4.2, we have the asymptotic null distribution of $T_{\boldsymbol{\alpha},\mathrm{GSXT},n}$.

Theorem 5.1 *Suppose Assumptions 5.1 and 5.2 hold. Under $H_{\boldsymbol{\alpha}}$, $T_{\boldsymbol{\alpha},\mathrm{GSXT},n}$ converges in distribution to the chi-square distribution with $r_{\boldsymbol{\alpha}}$ degrees of freedom as $\min_{\substack{i=1,\dots,a\\ j=1,\dots,b}} n_{ij} \to \infty$, where $r_{\boldsymbol{\alpha}} := \mathrm{rank}(\boldsymbol{V}_{\boldsymbol{\alpha}})$.*

By applying Theorem 5.1, we have, for a nominal level $\varphi \in (0, 1)$, an asymptotical size φ test if we reject H_{51} when $T_{\boldsymbol{\alpha},\mathrm{GSXT},n} \geq \chi^2_{\hat{r}_{n,\boldsymbol{\alpha}}}[1-\varphi]$, where $\hat{r}_{n,\boldsymbol{\alpha}} := \mathrm{rank}\left(\hat{\boldsymbol{V}}_{n,\boldsymbol{\alpha}}\right)$ and $\chi^2_{\hat{r}_{n,\boldsymbol{\alpha}}}[1-\varphi]$ denotes the upper φ-percentiles of the chi-square distribution with $\hat{r}_{n,\boldsymbol{\alpha}}$ degrees of freedom. The consistency of the test is shown in the following theorem.

Theorem 5.2 *Suppose Assumptions 5.1 and 5.2 hold. The proposed test based on $T_{n,\boldsymbol{\alpha}}$ is consistent, i.e., it holds that, under K_{51},*

$$\mathrm{P}\left(T_{\boldsymbol{\alpha},\mathrm{GSXT},n} \geq \chi^2_{\hat{r}_{n,\boldsymbol{\alpha}}}[1-\varphi]\right) \to 1 \quad as \quad \min_{\substack{i=1,\dots,a\\ j=1,\dots,b}} n_{ij} \to \infty.$$

***Proof** (**Theorem** 5.2)* From Theorem 2.1, we know

$$\left(\boldsymbol{e}_{1..}{}^\top - \boldsymbol{e}_{...}{}^\top \dots \boldsymbol{e}_{a..}{}^\top - \boldsymbol{e}_{...}{}^\top\right)^\top = O_p(1/\sqrt{n}).$$

Under $K_{\boldsymbol{\alpha}}$, we observe, for any $i_1, i_2 \in \{1, \dots, a\}$,

$$\begin{aligned} &\mathrm{P}\left(T_{n,\boldsymbol{\alpha}} \geq \chi^2_{\hat{r}_{n,\boldsymbol{\alpha}}}[1-\varphi]\right) \\ =&\mathrm{P}\left(T_{n,\boldsymbol{\alpha}}/n \geq \chi^2_{r_{\boldsymbol{\alpha}}}[1-\varphi]/n\right) + o_p(1) \\ =&\mathrm{P}\left(\begin{pmatrix} \boldsymbol{\alpha}_1 - \boldsymbol{\alpha}_. \\ \dots \\ \boldsymbol{\alpha}_a - \boldsymbol{\alpha}_. \end{pmatrix} \boldsymbol{V}^-_{\boldsymbol{\alpha}} \begin{pmatrix} \boldsymbol{\alpha}_1 - \boldsymbol{\alpha}_. \\ \dots \\ \boldsymbol{\alpha}_a - \boldsymbol{\alpha}_. \end{pmatrix}^\top \geq 0\right) + o_p(1), \end{aligned}$$

which tends to one as $\min_{\substack{i=1,\dots,a\\ j=1,\dots,b}} n_{ij} \to \infty$ since the Moore–Penrose inverse of any non-negative definite matrix is non-negative definite (Wu, 1980, Theorem 1). □

Next, we derive the nontrivial power under the local alternative. For any ap-by-ap symmetric, positive definite matrix ${}^{\alpha}\boldsymbol{H} = ({}^{\alpha}\boldsymbol{H}_{i_1 i_2})_{i_1,i_2=1\dots,a}$ with a p-by-p matrix ${}^{\alpha}\boldsymbol{H}_{i_1 i_2}$, the local alternative is defined as

$$K^{(n)}_{51}\ {}^{\alpha}\boldsymbol{\Sigma} := \frac{{}^{\alpha}\boldsymbol{H}}{n}.$$

Theorem 5.3 *Suppose Assumptions 5.1 and 5.2 hold. Under $K^{(n)}_{51}$, $T_{\boldsymbol{\alpha},\mathrm{GSXT},n}$ converges in distribution to $\boldsymbol{G}_{\boldsymbol{\alpha}} \boldsymbol{V}^-_{\boldsymbol{\alpha}} \boldsymbol{G}_{\boldsymbol{\alpha}}$ as $\min_{\substack{i=1,\dots,a\\ j=1,\dots,b}} n_{ij} \to \infty$, where $\boldsymbol{Z}_{\boldsymbol{\alpha}}$ follows an ap-dimensional centered normal distribution with variance ${}^{\alpha}\tilde{\boldsymbol{H}} + \boldsymbol{V}_{\boldsymbol{\alpha}}$ with ${}^{\alpha}\tilde{\boldsymbol{H}} = ({}^{\alpha}\tilde{\boldsymbol{H}}_{i_1 i_2})_{i_1,i_2=1\dots,a}$ defined as*

$$^{\alpha}\tilde{\boldsymbol{H}}_{i_1 i_2} = {}^{\alpha}\boldsymbol{H}_{i_1 i_2} - \frac{1}{a}\sum_{s=1}^{a}({}^{\alpha}\boldsymbol{H}_{i_1 s} + {}^{\alpha}\boldsymbol{H}_{s i_2}) + \frac{1}{a^2}\sum_{s_1, s_2=1}^{a} {}^{\alpha}\boldsymbol{H}_{s_1 s_2}.$$

Therefore, for a nominal level $\varphi \in (0, 1)$ *and under* $K_{51}^{(n)}$, *the local power of the test based on* $T_{\boldsymbol{\alpha},\mathrm{GSXT},n}$ *is given by*

$$\mathbb{P}\left(T_{\boldsymbol{\alpha},\mathrm{GSXT},n} \geq \chi^2_{\hat{r}_{n,\boldsymbol{\alpha}}}[1-\varphi]\right) \to \mathbb{P}\left(\boldsymbol{G}_{\boldsymbol{\alpha}} \boldsymbol{V}_{\boldsymbol{\alpha}}^{-} \boldsymbol{G}_{\boldsymbol{\alpha}} \geq \chi^2_{r_{\boldsymbol{\alpha}}}[1-\varphi]\right)$$

as $\min_{\substack{i=1,\ldots,a \\ j=1,\ldots,b}} n_{ij} \to \infty$.

Proof ***(Theorem*** 5.3*)* The proof can be completed by an application of the continuous mapping theorem. We can show that, under $K_{\boldsymbol{\alpha}}^{(n)}$,

$$\sqrt{n}\begin{pmatrix} \boldsymbol{Z}_{1..} - \boldsymbol{Z}_{...} \\ \vdots \\ \boldsymbol{Z}_{a..} - \boldsymbol{Z}_{...} \end{pmatrix} \Rightarrow N(\boldsymbol{0}, {}^{\alpha}\tilde{\boldsymbol{H}} + \boldsymbol{V}_{\boldsymbol{\alpha}}) \quad \text{as} \quad \min_{\substack{i=1,\ldots,a \\ j=1,\ldots,b}} n_{ij} \to \infty.$$

First observe that $\sqrt{n}\mathrm{E}(\boldsymbol{Z}_{i..} - \boldsymbol{Z}_{...}) = \boldsymbol{0}$ for any $i \in \{1, \ldots, a\}$. Next, for any $i_1, i_2 \in \{1, \ldots, a\}$, it follows that

$$\begin{aligned} & n\mathrm{Cov}(\boldsymbol{Z}_{i_1..} - \boldsymbol{Z}_{...}, \boldsymbol{Z}_{i_2..} - \boldsymbol{Z}_{...}) \\ =& n\mathrm{Cov}(\boldsymbol{\alpha}_{i_1} - \boldsymbol{\alpha}_{.}, \boldsymbol{\alpha}_{i_2} - \boldsymbol{\alpha}_{.}) + n\mathrm{Cov}(\boldsymbol{e}_{i_1..} - \boldsymbol{e}_{...}, \boldsymbol{e}_{i_2..} - \boldsymbol{e}_{...}) \\ =& {}^{\alpha}\tilde{\boldsymbol{H}}_{i_1 i_2} + {}^{n,\alpha}\boldsymbol{V}_{i_1 i_2}. \end{aligned}$$

By the Gaussian assumption of $\{\boldsymbol{\alpha}_i\}$ and (5.4), the cumulants of order three or greater for $\{\boldsymbol{\alpha}_i\}$ and $\{\boldsymbol{e}_{ijt}\}$ vanish asymptotically. □

Remark 5.4 The Gaussian assumption on $\boldsymbol{\alpha}_i$ can be relaxed. Let $(\boldsymbol{\alpha}_1^\top, \ldots, \boldsymbol{\alpha}_a^\top)^\top$ follow an ap-dimensional random vector. For the non-Gaussian random effects, the null and alternative hypotheses can be formulated as

$$H_{52} : \mathrm{P}(\boldsymbol{\alpha}_1 = \cdots = \boldsymbol{\alpha}_a) = 1 \quad \text{and} \quad K_{52} : \mathrm{P}(\boldsymbol{\alpha}_1 = \cdots = \boldsymbol{\alpha}_a) = 0.$$

We can show that the test based on $T_{\boldsymbol{\alpha},\mathrm{GSXT},n}$ has the asymptotic size φ and is consistent in the same manner as Theorems 5.1 and 5.2. Moreover, under the local alternative $K_{\boldsymbol{\alpha}}$, defined as

$$K_{52}^{(n)} \ \boldsymbol{\alpha}_i := \frac{\boldsymbol{h}_i}{\sqrt{n}},$$

where $(\boldsymbol{h}_1^\top, \ldots, \boldsymbol{h}_a^\top)^\top$ follows an ap-dimensional centered random vector, it holds that, for $\boldsymbol{h}_{.} := \sum_{i=1}^{a} \boldsymbol{h}_i / a$,

$$T_{\boldsymbol{\alpha},\mathrm{GSXT},n} \Rightarrow \left(\begin{pmatrix} \boldsymbol{h}_1 - \boldsymbol{h}_. \\ \vdots \\ \boldsymbol{h}_a - \boldsymbol{h}_. \end{pmatrix} + \boldsymbol{Z}_{\boldsymbol{\alpha}}\right)^\top \boldsymbol{V}_{\boldsymbol{\alpha}}^- \left(\begin{pmatrix} \boldsymbol{h}_1 - \boldsymbol{h}_. \\ \vdots \\ \boldsymbol{h}_a - \boldsymbol{h}_. \end{pmatrix} + \boldsymbol{Z}_{\boldsymbol{\alpha}}\right) \quad \text{as} \min_{\substack{i=1,\ldots,a \\ j=1,\ldots,b}} n_{ij} \to \infty,$$

where $\boldsymbol{Z}_{\boldsymbol{\alpha}}$ follows an ap-dimensional centered normal distribution with variance $\boldsymbol{V}_{\boldsymbol{\alpha}}$.

Remark 5.5 Fixed effect models are the same as (5.1) but $\{\boldsymbol{\alpha}_i\}$ and $\{\boldsymbol{\beta}_j\}$ are fixed effects, i.e., non-random constants, satisfying $\sum_{i=1}^a \boldsymbol{\alpha}_i = \boldsymbol{0}_p$ and $\sum_{j=1}^b \boldsymbol{\beta}_j = \boldsymbol{0}_p$. A test for the existence of fixed effects can be described as

$$\begin{aligned} &H_{53} : \boldsymbol{\alpha}_i = \boldsymbol{0} \text{ for all } i \in \{1, \ldots, a\} \\ \text{vs} \quad &K_{53} : \boldsymbol{\alpha}_i \neq \boldsymbol{0} \quad \text{for some } i \in \{1, \ldots, a\}. \end{aligned}$$

We can utilize the same test based on $T_{n,\boldsymbol{\alpha},\mathrm{GSXT}}$. Under H_{53}, the fixed effect model is exactly the same model with the random effect model under H_{51} and, thus, Theorem 5.1 holds. The consistency of the test follows along the lines of the proof of Theorem 5.2. The difference between fixed and random effects shows up in the local power. The local alternative is defined, for constants $\{{}^{\alpha}\boldsymbol{h}_i; i = 1, \ldots, a\}$ such that $\sum_{i=1}^a {}^{\alpha}\boldsymbol{h}_i = \boldsymbol{0}$, as

$$K_{53}^{(n)} \ \boldsymbol{\alpha}_i := \frac{{}^{\alpha}\boldsymbol{h}_i}{\sqrt{n}}.$$

We can apply Lemma 4.2 to deduce that, under $K_{53}^{(n)}$, $T_{\boldsymbol{\alpha},\mathrm{GSXT},n}$ converges in distribution to the noncentral chi-square distribution with $r_{\boldsymbol{\alpha}}$ degrees of freedom and the noncentrality parameter $\delta_\alpha := ({}^{\alpha}\boldsymbol{h}_1^\top, \ldots, {}^{\alpha}\boldsymbol{h}_a^\top)\boldsymbol{V}_{\boldsymbol{\alpha}}^-({}^{\alpha}\boldsymbol{h}_1^\top, \ldots, {}^{\alpha}\boldsymbol{h}_a^\top)^\top$ as $\min_{\substack{i=1,\ldots,a \\ j=1,\ldots,b}} n_{ij} \to \infty$, where $r_{\boldsymbol{\alpha}} := \mathrm{rank}(\boldsymbol{V}_{\boldsymbol{\alpha}})$. Hence, the local power of the test is given by $1 - \Psi_{r_{\boldsymbol{\alpha}},\delta_\alpha}(\chi^2_{r_{\boldsymbol{\alpha}}}[1-\tau])$, where $\Psi_{r_{\boldsymbol{\alpha}},\delta_\alpha}$ is the cumulative distribution function of the noncentral chi-square distribution with $r_{\boldsymbol{\alpha}}$ degrees of freedom and the noncentrality parameter δ_α.

Remark 5.6 Since $\boldsymbol{Z}_{i..} - \boldsymbol{Z}_{...} = \boldsymbol{\alpha}_i - \boldsymbol{\alpha}_. + \boldsymbol{e}_{i..} - \boldsymbol{e}_{...}$, $T_{n,\boldsymbol{\alpha}}$ is independent of $\{\boldsymbol{\beta}_j\}$. Thereby, we can perform the test for the hypothesis (5.3) through the following test statistic:

$$T_{\boldsymbol{\beta},\mathrm{GSXT},n} := \begin{pmatrix} \boldsymbol{Z}_{.1.} - \boldsymbol{Z}_{...} \\ \vdots \\ \boldsymbol{Z}_{.a.} - \boldsymbol{Z}_{...} \end{pmatrix} \hat{\boldsymbol{V}}_{n,\boldsymbol{\beta}}^- \begin{pmatrix} \boldsymbol{Z}_{.1.} - \boldsymbol{Z}_{...} \\ \vdots \\ \boldsymbol{Z}_{.a.} - \boldsymbol{Z}_{...} \end{pmatrix}^\top,$$

where $\hat{\boldsymbol{V}}_{n,\boldsymbol{\beta}} := \hat{\boldsymbol{V}}_{n,\boldsymbol{\alpha}}$.

Remark 5.7 The mixed effect model is the same as (5.1) but $\{\boldsymbol{\alpha}_i\}$ is a fixed effect and $\{\boldsymbol{\beta}_j\}$ is a random effect. By Remark 5.6, we can test the existence of $\{\boldsymbol{\alpha}_i\}$ and $\{\boldsymbol{\beta}_j\}$ by the tests based on $T_{\boldsymbol{\alpha},\mathrm{GSXT},n}$ and $T_{\boldsymbol{\beta},\mathrm{GSXT},n}$, respectively.

5.2 Test for the Existence of Random Interactions

The two-way random effect model with random interactions, time-dependent errors, and correlated groups is defined as

$$\boldsymbol{Z}_{ijt} = \boldsymbol{\mu} + \boldsymbol{\alpha}_i + \boldsymbol{\beta}_j + \boldsymbol{\gamma}_{ij} + \boldsymbol{e}_{ijt}, \quad i = 1, \ldots, a;\ j = 1, \ldots, b;\ t = 1, \ldots, n_{ij}, \tag{5.6}$$

where $\boldsymbol{\gamma}_{ij} := (\gamma_{ij1}, \ldots, \gamma_{ijp})^\top$ is an interaction between the i-th level of factor A and the j-th level of factor B, and the other terms play the same role as (5.1).

We consider Assumptions (i), (iii), and (iv) described below (5.1). Instead of (ii), we impose that

(ii′) any two of $\{\boldsymbol{\alpha}_i\}$, $\{\boldsymbol{\beta}_j\}$, $\{\boldsymbol{\gamma}_{ij}\}$, and $\{\boldsymbol{e}_{ijt}\}$ are independent.

For the sake of simplicity, we assume that

$$(\boldsymbol{\gamma}_{11}^\top, \boldsymbol{\gamma}_{21}^\top, \ldots, \boldsymbol{\gamma}_{a1}^\top, \boldsymbol{\gamma}_{12}^\top, \ldots, \boldsymbol{\gamma}_{a2}^\top, \ldots, \boldsymbol{\gamma}_{1b}^\top, \ldots, \boldsymbol{\gamma}_{ab}^\top)^\top$$

follows an abp-dimensional centered normal distribution with variance ${}^\gamma\boldsymbol{\Sigma} := ({}^\gamma\boldsymbol{\Sigma}_{j_1 j_2})_{j_1, j_2 = 1, \ldots, b}$, where ${}^\gamma\boldsymbol{\Sigma}_{j_1 j_2} := ({}^\gamma\boldsymbol{\Sigma}_{j_1 j_2}^{i_1 i_2})_{i_1, i_2 = 1, \ldots, a}$ and ${}^\gamma\boldsymbol{\Sigma}_{j_1 j_2}^{i_1 i_2} := \mathrm{E}(\boldsymbol{\gamma}_{i_1 j_1} \boldsymbol{\gamma}_{i_2 j_2}^\top)$.

A test for the existence of interaction effects can be formulated as follows:

$$H_{54} : {}^\gamma\boldsymbol{\Sigma} = \boldsymbol{O}_{abp} \quad \text{vs} \quad K_{54} : {}^\gamma\boldsymbol{\Sigma} \neq \boldsymbol{O}_{abp}. \tag{5.7}$$

Next, we briefly explain the motivation of the test statistic. The elementary calculation gives the following decomposition into sum of squares:

$$\begin{aligned}
&\sum_{i=1}^{a}\sum_{j=1}^{b}\sum_{t=1}^{n_{ij}} (\mathbf{Z}_{ijt} - \mathbf{Z}_{...})^\top (\mathbf{Z}_{ijt} - \mathbf{Z}_{...}) \\
=&\sum_{i=1}^{a}\sum_{j=1}^{b}\sum_{t=1}^{n_{ij}} \Big\{ (\mathbf{Z}_{ijt} - \mathbf{Z}_{ij.}) + (\mathbf{Z}_{ij.} - \mathbf{Z}_{i..} - \mathbf{Z}_{.j.} + \mathbf{Z}_{...}) \\
&+ (\mathbf{Z}_{i..} - \mathbf{Z}_{...}) + (\mathbf{Z}_{.j.} - \mathbf{Z}_{...}) \Big\}^\top \\
&\times \left\{(\mathbf{Z}_{ijt} - \mathbf{Z}_{ij.}) + (\mathbf{Z}_{ij.} - \mathbf{Z}_{i..} - \mathbf{Z}_{.j.} + \mathbf{Z}_{...}) + (\mathbf{Z}_{i..} - \mathbf{Z}_{...}) + (\mathbf{Z}_{.j.} - \mathbf{Z}_{...})\right\} \\
=&\sum_{i=1}^{a}\sum_{j=1}^{b}\sum_{t=1}^{n_{ij}} \Big\{ (\mathbf{Z}_{ijt} - \mathbf{Z}_{ij.})^\top (\mathbf{Z}_{ijt} - \mathbf{Z}_{ij.}) + (\mathbf{Z}_{i..} - \mathbf{Z}_{...})^\top (\mathbf{Z}_{i..} - \mathbf{Z}_{...}) \\
&+ (\mathbf{Z}_{ij.} - \mathbf{Z}_{i..} - \mathbf{Z}_{.j.} + \mathbf{Z}_{...})^\top (\mathbf{Z}_{ij.} - \mathbf{Z}_{i..} - \mathbf{Z}_{.j.} + \mathbf{Z}_{...}) \\
&+ (\mathbf{Z}_{.j.} - \mathbf{Z}_{...})^\top (\mathbf{Z}_{.j.} - \mathbf{Z}_{...}) \Big\}.
\end{aligned}$$

See detailed discussion in Clarke (2008, Section 5.6). The term

$$\left(\boldsymbol{Z}_{ij.} - \boldsymbol{Z}_{i..} - \boldsymbol{Z}_{.j.} + \boldsymbol{Z}_{...}\right)^\top \left(\boldsymbol{Z}_{ij.} - \boldsymbol{Z}_{i..} - \boldsymbol{Z}_{.j.} + \boldsymbol{Z}_{...}\right)$$

corresponds to the sum of squares for the interaction, and we shall apply a similar idea as in Sections 4 and 5.2 to $\boldsymbol{Z}_{ij.} - \boldsymbol{Z}_{i..} - \boldsymbol{Z}_{.j.} + \boldsymbol{Z}_{...}$.

From the above discussion, a test statistic for the hypothesis (5.7) is defined as

$$T_{\gamma,\mathrm{GSXT},n} := \begin{pmatrix} \boldsymbol{Z}_{11.} - \boldsymbol{Z}_{1..} - \boldsymbol{Z}_{.1.} + \boldsymbol{Z}_{...} \\ \boldsymbol{Z}_{21.} - \boldsymbol{Z}_{2..} - \boldsymbol{Z}_{.1.} + \boldsymbol{Z}_{...} \\ \vdots \\ \boldsymbol{Z}_{a1.} - \boldsymbol{Z}_{a..} - \boldsymbol{Z}_{.1.} + \boldsymbol{Z}_{...} \\ \boldsymbol{Z}_{12.} - \boldsymbol{Z}_{1..} - \boldsymbol{Z}_{.2.} + \boldsymbol{Z}_{...} \\ \vdots \\ \boldsymbol{Z}_{a2.} - \boldsymbol{Z}_{a..} - \boldsymbol{Z}_{.2.} + \boldsymbol{Z}_{...} \\ \vdots \\ \boldsymbol{Z}_{1b.} - \boldsymbol{Z}_{1..} - \boldsymbol{Z}_{.b.} + \boldsymbol{Z}_{...} \\ \vdots \\ \boldsymbol{Z}_{ab.} - \boldsymbol{Z}_{a..} - \boldsymbol{Z}_{.b.} + \boldsymbol{Z}_{...} \end{pmatrix}^\top \hat{\boldsymbol{V}}_{n,\gamma}^- \begin{pmatrix} \boldsymbol{Z}_{11.} - \boldsymbol{Z}_{1..} - \boldsymbol{Z}_{.1.} + \boldsymbol{Z}_{...} \\ \boldsymbol{Z}_{21.} - \boldsymbol{Z}_{2..} - \boldsymbol{Z}_{.1.} + \boldsymbol{Z}_{...} \\ \vdots \\ \boldsymbol{Z}_{a1.} - \boldsymbol{Z}_{a..} - \boldsymbol{Z}_{.1.} + \boldsymbol{Z}_{...} \\ \boldsymbol{Z}_{12.} - \boldsymbol{Z}_{1..} - \boldsymbol{Z}_{.2.} + \boldsymbol{Z}_{...} \\ \vdots \\ \boldsymbol{Z}_{a2.} - \boldsymbol{Z}_{a..} - \boldsymbol{Z}_{.2.} + \boldsymbol{Z}_{...} \\ \vdots \\ \boldsymbol{Z}_{1b.} - \boldsymbol{Z}_{1..} - \boldsymbol{Z}_{.b.} + \boldsymbol{Z}_{...} \\ \vdots \\ \boldsymbol{Z}_{ab.} - \boldsymbol{Z}_{a..} - \boldsymbol{Z}_{.b.} + \boldsymbol{Z}_{...} \end{pmatrix}, \tag{5.8}$$

where $\hat{\boldsymbol{V}}_{n,\gamma}^-$ is the Moore–Penrose inverse matrix of $\hat{\boldsymbol{V}}_{n,\gamma} := ({}^{\gamma}\hat{\boldsymbol{V}}_{j_1 j_2})_{j_1, j_2=1,\dots,b}$, ${}^{\gamma}\hat{\boldsymbol{V}}_{j_1 j_2} := ({}^{\gamma}\hat{\boldsymbol{V}}^{i_1 i_2}_{j_1 j_2})_{i_1, i_2=1,\dots,a}$, ${}^{\gamma}\hat{\boldsymbol{V}}^{i_1 i_2}_{j_1 j_2} := \sum_{v=1}^{4} \hat{\boldsymbol{\xi}}_v$, $\hat{\boldsymbol{\xi}}_0 := \hat{\boldsymbol{\zeta}}^{i_1 i_2}_{j_1 j_2}$, $\hat{\boldsymbol{\xi}}_1 := -\hat{\boldsymbol{\zeta}}^{.i_2}_{j_1 j_2} - \hat{\boldsymbol{\zeta}}^{i_1 .}_{j_1 j_2} - \hat{\boldsymbol{\zeta}}^{i_1 i_2}_{.j_2} - \hat{\boldsymbol{\zeta}}^{i_1 i_2}_{j_1 .}$, $\hat{\boldsymbol{\xi}}_2 := \hat{\boldsymbol{\zeta}}^{..}_{j_1 j_2} + \hat{\boldsymbol{\zeta}}^{.i_2}_{.j_2} + \hat{\boldsymbol{\zeta}}^{.i_2}_{j_1 .} + \hat{\boldsymbol{\zeta}}^{i_1 .}_{.j_2} + \hat{\boldsymbol{\zeta}}^{i_1 .}_{j_1 .} + \hat{\boldsymbol{\zeta}}^{i_1 i_2}_{..}$, $\hat{\boldsymbol{\xi}}_3 := -\hat{\boldsymbol{\zeta}}^{i_1 .}_{..} - \hat{\boldsymbol{\zeta}}^{.i_2}_{..} - \hat{\boldsymbol{\zeta}}^{..}_{j_1 .} - \hat{\boldsymbol{\zeta}}^{..}_{.j_2}$, $\hat{\boldsymbol{\xi}}_4 := \hat{\boldsymbol{\zeta}}^{..}_{..}$, and a subscript dot denotes taking the average with respect to the corresponding element. As with $T_{\boldsymbol{\alpha},\mathrm{GSXT},n}$, $\boldsymbol{V}_\gamma$ is a singular matrix since $\sum_{i_1=1}^{a} \sum_{j_1=1}^{b} {}^{\gamma}\boldsymbol{V}^{i_1 i_2}_{j_1 j_2} = \boldsymbol{O}_p$, and, thus, the Moore–Penrose inverse is necessary.

To ensure the consistency of $\hat{\boldsymbol{V}}_{n,\gamma}^-$, we make the following assumption.

Assumption 5.3 It holds that $\mathrm{rank}(\hat{\boldsymbol{V}}_{n,\gamma})$ converges in probability to $\mathrm{rank}(\boldsymbol{V}_\gamma)$ as $\min_{\substack{i=1,\dots,a \\ j=1,\dots,b}} n_{ij} \to \infty$.

Then, we obtain the asymptotic null distribution of $T_{\gamma,\mathrm{GSXT},n}$.

Theorem 5.4 *Suppose Assumptions 5.1, 5.2 (i), and 5.3 hold. Under* H_{54}, $T_{\gamma,\mathrm{GSXT},n}$ *converges in distribution to the chi-square distribution with* r_γ *degrees of freedom as* $\min_{\substack{i=1,\dots,a \\ j=1,\dots,b}} n_{ij} \to \infty$, *where* $r_\gamma := \mathrm{rank}(\boldsymbol{V}_\gamma)$ *and* $\boldsymbol{V}_\gamma$ *is defined by replacing* $\hat{\boldsymbol{f}}^{i_1 i_2}_{j_1 j_2}(0)$ *with* $\boldsymbol{f}^{i_1 i_2}_{j_1 j_2}(0)$ *in* $\hat{\boldsymbol{V}}_{n,\gamma}$.

***Proof (Theorem** 5.4)* We omit the proof since it is analogous to the proof of Theorem 5.4. □

According to Theorem 5.4, we can derive a test with asymptotic size φ if we reject H_{54} when $T_{\gamma,\mathrm{GSXT},n} \geq \chi^2_{\hat{r}_{n,\gamma}}[1-\varphi]$, where $\hat{r}_{n,\gamma} := \mathrm{rank}\left(\hat{\boldsymbol{V}}_{n,\gamma}\right)$. The consistency of the test is shown along the lines of Theorem 5.2.

Theorem 5.5 *Suppose Assumptions 5.1, 5.2 (i), and 5.3 hold. The test based on $T_{n,\gamma}$ is consistent, i.e., it holds that, under K_{γ},*

$$\mathrm{P}\left(T_{\gamma,\mathrm{GSXT},n} \geq \chi^2_{\hat{r}_{n,\gamma}}[1-\varphi]\right) \to 1 \quad as \quad \min_{\substack{i=1,\dots,a\\ j=1,\dots,b}} n_{ij} \to \infty.$$

***Proof** (Theorem 5.5)* We omit the proof since it is analogous to the proof of Theorem 5.5. □

Let $K_{55}^{(n)}$ be the local alternative

$$K_{55}^{(n)} := \frac{{}^{\gamma}\boldsymbol{H}}{n},$$

where abp-by-abp symmetric, positive definite matrix ${}^{\gamma}\boldsymbol{H} = ({}^{\gamma}\boldsymbol{H}_{j_1 j_2})_{j_1,j_2=1\dots,b}$ with a ap-by-ap matrix ${}^{\gamma}\boldsymbol{H}_{j_1 j_2} := ({}^{\gamma}\boldsymbol{H}^{i_1 i_2}_{j_1 j_2})_{i_1,i_2=1,\dots,a}$.

Theorem 5.6 *Suppose Assumptions 5.1, 5.2 (i), and 5.3 hold. Then, the local power of the test based on $T_{\gamma,\mathrm{GSXT},n}$ under $K_{\gamma}^{(n)}$ is given by*

$$\mathbb{P}(T_{\gamma,\mathrm{GSXT},n} \geq \chi^2_{\hat{r}_{n,\gamma}}[1-\varphi]) \to \mathbb{P}(\boldsymbol{G}_{\gamma}\boldsymbol{V}_{\gamma}^{-}\boldsymbol{G}_{\gamma} \geq \chi^2_{r_{\gamma}}[1-\varphi]) \; as \; \min_{\substack{i=1,\dots,a\\ j=1,\dots,b}} n_{ij} \to \infty,$$

where $\boldsymbol{G}_{\gamma}$ is an abp-dimensional centered normal random variable with variance ${}^{\gamma}\tilde{\boldsymbol{H}} + \boldsymbol{V}_{\gamma}$ and ${}^{\gamma}\tilde{\boldsymbol{H}} = ({}^{\gamma}\tilde{\boldsymbol{H}}_{i_1 i_2})_{i_1,i_2=1\dots,a}$ is defined by replacing $\hat{\boldsymbol{\zeta}}^{i_1 i_2}_{j_1 j_2}$ with ${}^{\gamma}\boldsymbol{H}^{i_1 i_2}_{j_1 j_2}$ in $\hat{\boldsymbol{V}}_{n,\gamma}$.

***Proof** (Theorem 5.6)* We omit the proof since it is analogous to the proof of Theorem 5.6. □

Remark 5.8 Similar to Remark 5.4, Gaussianity of interactions can be relaxed.

Remark 5.9 The case that the interaction $\{\boldsymbol{\gamma}_{ij}\}$ is a non-random constant can be dealt with in the same manner. See also Remark 5.5.

Remark 5.10 Since

$$\boldsymbol{Z}_{ij.} - \boldsymbol{Z}_{i..} - \boldsymbol{Z}_{.j.} + \boldsymbol{Z}_{...} = \boldsymbol{\gamma}_{ij} - \boldsymbol{\gamma}_{i.} - \boldsymbol{\gamma}_{.j} + \boldsymbol{\gamma}_{..} + \boldsymbol{e}_{ij.} - \boldsymbol{e}_{i..} - \boldsymbol{e}_{.j.} + \boldsymbol{e}_{...},$$

$T_{n,\gamma}$ does not depend on $\{\boldsymbol{\alpha}_i\}$ and $\{\boldsymbol{\beta}_j\}$. Thus, $\boldsymbol{\alpha}_i$ and $\boldsymbol{\beta}_j$ can be either fixed or random effects.

Remark 5.11 The test for the existence of fixed (or random) effects $\{\boldsymbol{\alpha}_i\}$ for the two-way models with interactions (5.6) cannot be constructed by the quantity $\boldsymbol{Z}_{i..} - \boldsymbol{Z}_{...}$ since $\boldsymbol{Z}_{i..} - \boldsymbol{Z}_{...} = \boldsymbol{\alpha}_i - \boldsymbol{\alpha}_{.} + \boldsymbol{\gamma}_{i.} - \boldsymbol{\gamma}_{..} + \boldsymbol{e}_{i..} - \boldsymbol{e}_{...}$, and, thus, $\sqrt{n}\,(\boldsymbol{Z}_{i..} - \boldsymbol{Z}_{...}) = O_p(\sqrt{n})$ under the null hypothesis $\boldsymbol{\alpha}_1 = \cdots = \boldsymbol{\alpha}_a$ unless $\boldsymbol{\gamma}_{i.} - \boldsymbol{\gamma}_{..} = 0$.

References

Brillinger, D. R. (1981). *Time series: Data analysis and theory*. San Francisco: Holden-Day.

Clarke, B. R. (2008). *Linear models: The theory and application of analysis of variance*. Wiley.

Goto, Y., Suzuki, K., Xu, X., & Taniguchi, M. (2023). Tests for the existence of group effects and interactions for two-way models with dependent errors. *Annals of the Institute of Statistical Mathematics, 75,* 511–532.

Rakocevic, V. (1997). On continuity of the Moore-Penrose and Drazin inverses. *Matematicki Vesnik, 49,* 163–172.

Wu, C.-F. (1980). On some ordering properties of the generalized inverses of nonnegative definite matrices. *Linear Algebra and Its Applications, 32,* 49–60.

Chapter 6
Optimal Test for One-Way Random Effect Model

The optimality of the tests has not been discussed so far. For the test for the existence of fixed effects, the locally asymptotically maximin test based on local asymptotic normality (LAN) was proposed by Hallin et al. (2021) for i.i.d. data. In this chapter, we show that the one-way random effect model for i.i.d. sequences does not have the LAN property, and, thus, the LAN framework cannot be applied to construct the optimal test. We introduce the one-way random effect model and the likelihood ratio process in Section 6.1. With the aid of the Neyman–Pearson lemma, we reveal that the likelihood ratio test is asymptotically most powerful for the contiguous hypothesis that the variance of random effects belongs to the boundary of parameter space under the null in Section 6.2. On the other hand, we prove that the likelihood ratio process converges to zero under the null when the contiguous hypothesis that the variance of the random effects belongs to the interior of parameter space under the null hypothesis in Section 6.3. This chapter follows mainly the development of Goto et al. (2022).

6.1 Log-Likelihood Ratio Process for One-Way Random Effect Model

We consider the one-way random effect model : for $i = 1, \ldots, a$ and $t = 1, \ldots, n$,

$$Z_{it} = \mu + \alpha_i + e_{it}, \tag{6.1}$$

where Z_{it} is a t-th one-dimensional observation from an i-th group, μ is a one-dimensional general mean, α_i is a random effect, and e_{it} is a one-dimensional time series of i-th group at time t. Equivalently, it can be rewritten in matrix form:

$$\boldsymbol{Z} = \mu \mathbf{1}_{an} + (\mathbf{1}_a \otimes \mathbf{1}_n)\boldsymbol{\alpha} + \boldsymbol{e},$$

Y. Goto et al., *ANOVA with Dependent Errors*, JSS Research Series in Statistics,
https://doi.org/10.1007/978-981-99-4172-8_6

where $\boldsymbol{Z} := (Z_{11}, \dots, Z_{1n}, Z_{21}, \dots, Z_{a1}, \dots, Z_{an})^\top$, $\boldsymbol{\alpha} := (\alpha_1, \dots, \alpha_a)^\top$, and $\boldsymbol{e} := (e_{11}, \dots, e_{1n}, e_{21}, \dots\dots, e_{a1}, \dots, e_{an})^\top$.

We suppose $(\alpha_i, e_{i'j})^\top$ follows the i.i.d. centered normal distribution with variance $\begin{pmatrix} \sigma_\alpha^2 & 0 \\ 0 & \sigma_e^2 \end{pmatrix}$ for any $i, i' = 1, \dots, a$ and $j = 1, \dots, n$.

By $\boldsymbol{Z} \sim N(\mu \mathbf{1}_{an}, I_a \otimes (\sigma_\alpha^2 J_n + \sigma_e^2 I_n))$ (see Searle et al., 1992, p.79), we know that the log-likelihood function $L_Z(\sigma_\alpha^2, \sigma_e^2, \mu)$ of $\boldsymbol{Z}$ is given by

$$\begin{aligned} L_Z(\sigma_\alpha^2, \sigma_e^2, \mu) := & -\frac{an}{2} \log 2\pi - \frac{a(n-1)}{2} \log \sigma_e^2 - \frac{a}{2} \log\left(\sigma_e^2 + n\sigma_\alpha^2\right) \\ & - \frac{1}{2\sigma_e^2} \sum_{i=1}^{a} \sum_{t=1}^{n} (Z_{it} - \mu)^2 + \frac{n^2 \sigma_\alpha^2}{2\sigma_e^2 \left(\sigma_e^2 + n\sigma_\alpha^2\right)} \sum_{i=1}^{a} (Z_{i\cdot} - \mu)^2, \end{aligned} \tag{6.2}$$

where $Z_{i\cdot} := \sum_{t=1}^{n} Z_{it}/n$. For the sake of notational convenience, let us define $\boldsymbol{\theta} = (\theta_1, \theta_2, \theta_3) := (\sigma_\alpha^2, \sigma_e^2, \mu)$ and the log-likelihood ratio process, for the parameters $\boldsymbol{\theta}_0$ under the null hypothesis and $\boldsymbol{\theta}_n$ under the contiguous alternative, as

$$\Lambda(\boldsymbol{\theta}_0, \boldsymbol{\theta}_n) := L_Z(\boldsymbol{\theta}_n) - L_Z(\boldsymbol{\theta}_0). \tag{6.3}$$

6.2 The Hypothesis That the Variance of Random Effect Equals Zero Under the Null

In this section, we focus on the contiguous hypothesis that the variance of the random effects belongs to the boundary of parameter space under the null:

$$\mathrm{H}_{61} : \boldsymbol{\theta} = \boldsymbol{\theta}_{0,61} := \begin{pmatrix} 0 \\ \theta_2 \\ \theta_3 \end{pmatrix} \quad \text{vs} \quad \mathrm{K}_{61}^{(n)} : \boldsymbol{\theta} = \boldsymbol{\theta}_{n,61} := \begin{pmatrix} \frac{h_1}{n^{k_1}} \\ \theta_2 + \frac{h_2}{n^{k_2}} \\ \theta_3 + \frac{h_3}{n^{k_3}} \end{pmatrix},$$

where $k_1 > 0, k_2 > 0, k_3 > 0, \theta_2 > 0, h_1 > 0$, and $h_2 > -n^{k_2}\theta_2$.

Theorem 6.1 *(i) Under the null hypothesis* H_{61}*, the log-likelihood ratio* $\Lambda(\boldsymbol{\theta}_{0,61}, \boldsymbol{\theta}_{n,61})$ *has the following asymptotic expansion: for* $h \in \mathbb{R}$ *and sufficiently large* n *such that* $h_2 > -n^{k_2}\theta_2$*,*

$$\Lambda(\boldsymbol{\theta}_{0,61}, \boldsymbol{\theta}_{n,61}) = \begin{cases} \begin{pmatrix} \frac{h_3\sqrt{\theta}_2}{\theta_2 + h_1} & \frac{h_1}{2(\theta_2 + h_1)} \end{pmatrix} \begin{pmatrix} g_1(\boldsymbol{T}_{1,\mathrm{GKKT},n}) \\ g_2(\boldsymbol{T}_{1,\mathrm{GKKT},n}) \end{pmatrix} - \frac{a}{2}\log(1 + \frac{h_1}{\theta_2}) - \frac{ah_3^2}{2(\theta_2 + h_1)} + o_P(1) & k_2 \geq 1,\ k_3 = \frac{1}{2},\ k_1 = 1, \\ \frac{h_3}{\sqrt{\theta}_2} g_1(\boldsymbol{T}_{1,\mathrm{GKKT},n}) - \frac{ah_3^2}{2\theta_2} + o_P(1) & k_2 \geq 1,\ k_3 = \frac{1}{2},\ k_1 > 1, \\ \frac{h_1}{2(\theta_2 + h_1)} g_2(\boldsymbol{T}_{1,\mathrm{GKKT},n}) - \frac{a}{2}\log(1 + \frac{h_1}{\theta_2}) + o_P(1) & k_2 \geq 1,\ k_3 > \frac{1}{2},\ k_1 = 1, \\ o_P(1) & k_2 \geq 1,\ k_3 > \frac{1}{2},\ k_1 > 1, \end{cases}$$

where

$$\boldsymbol{T}_{1,\mathrm{GKKT},n} := \left(\frac{\sqrt{n}(Z_{1\cdot} - \theta_3)}{\sqrt{\theta_2}}, \ldots, \frac{\sqrt{n}(Z_{a\cdot} - \theta_3)}{\sqrt{\theta_2}} \right)^\top, \tag{6.4}$$

which follows, under $\mathrm{H}_{61}^{(n)}$, *the a-dimensional standard normal distribution, and*

$$\left(g_1\left((x_1, \ldots, x_a)^\top\right) \; g_2\left((x_1, \ldots, x_a)^\top\right)\right)^\top := \left(\sum_{i=1}^a x_i \; \sum_{i=1}^a x_i^2\right)^\top.$$

(ii) Under the null hypothesis $\mathrm{H}_{61}^{(n)}$, *the Fisher information of the model* (6.1) *is given by*

$$\mathcal{I}(\boldsymbol{\theta}_{0,61}) := \begin{pmatrix} \frac{1}{2(\theta_1+\theta_2)^2} & \frac{1}{2(\theta_1+\theta_2)^2} & 0 \\ \frac{1}{2(\theta_1+\theta_2)^2} & \frac{1}{2(\theta_1+\theta_2)^2} & 0 \\ 0 & 0 & \frac{1}{\theta_1+\theta_2} \end{pmatrix}.$$

Proof (Theorem 6.1*)* We only prove (i) since (ii) is an immediate consequence of the definition of the Fisher information matrix. Let the elements of $\boldsymbol{\theta}_{n,63}$ denote $\left(\theta_1^{(n)} \; \theta_2^{(n)} \; \theta_3^{(n)}\right)^\top$. From the definition of $\Lambda(\boldsymbol{\theta}_{0,61}, \boldsymbol{\theta}_{n,61})$, a lengthy but straightforward calculation gives

$$\begin{aligned}
&\Lambda(\boldsymbol{\theta}_{0,61}, \boldsymbol{\theta}_{n,61}) \\
&= \frac{n\left(\theta_3 - \theta_3^{(n)}\right)}{\theta_2^{(n)} + n\theta_1^{(n)}} \sum_{i=1}^a (Z_{i.} - \theta_3) - \frac{an\left(\theta_3^{(n)} - \theta_3\right)^2}{2\left(\theta_2^{(n)} + n\theta_1^{(n)}\right)} - \frac{a(n-1)}{2} \log \frac{\theta_2^{(n)}}{\theta_2} \\
&\quad + \frac{\theta_2^{(n)} - \theta_2}{2\theta_2^{(n)}\theta_2} \sum_{i=1}^a \sum_{t=1}^n (Z_{it} - Z_{i.})^2 - \frac{a}{2} \log \frac{\theta_2^{(n)} + n\theta_1^{(n)}}{\theta_2 + n\theta_1} \\
&\quad + \frac{n(\theta_2^{(n)} - \theta_2) + n^2\theta_1^{(n)}}{2\theta_2\left(\theta_2^{(n)} + n\theta_1^{(n)}\right)} \sum_{i=1}^a (Z_{i.} - \theta_3)^2 - \frac{n^2}{2} \frac{\theta_1}{\theta_2\left(\theta_2 + n\theta_1\right)} \sum_{i=1}^a (Z_{i\cdot} - \theta_3)^2 \\
&= L_1 + L_2 + L_3 + L_4,
\end{aligned}$$

where

$$L_1 := \frac{n\left(\theta_3 - \theta_3^{(n)}\right)}{\theta_2^{(n)} + n\theta_1^{(n)}} \sum_{i=1}^a (Z_{i.} - \theta_3), \quad L_2 := -\frac{an\left(\theta_3^{(n)} - \theta_3\right)^2}{2\left(\theta_2^{(n)} + n\theta_1^{(n)}\right)},$$

$$L_3 := -\frac{a(n-1)}{2} \log \frac{\theta_2^{(n)}}{\theta_2} + \frac{\theta_2^{(n)} - \theta_2}{2\theta_2^{(n)}\theta_2} \sum_{i=1}^a \sum_{t=1}^n (Z_{it} - Z_{i.})^2,$$

$$\text{and } L_4 := -\frac{a}{2}\log\frac{\theta_2^{(n)}+n\theta_1^{(n)}}{\theta_2+n\theta_1} + \frac{n(\theta_2^{(n)}-\theta_2)+n^2\theta_1^{(n)}}{2\theta_2\left(\theta_2^{(n)}+n\theta_1^{(n)}\right)}\sum_{i=1}^{a}(Z_{i.}-\theta_3)^2$$
$$-\frac{n^2}{2}\frac{\theta_1}{\theta_2\left(\theta_2+n\theta_1\right)}\sum_{i=1}^{a}(Z_{i.}-\theta_3)^2 .$$

We can show the following convergences: for $k_1, k_2, k_3 > 0$,

(i)

$$L_1 = -\frac{h_3}{n^{k_3-1}\left(\theta_2+\frac{h_2}{n^{k_2}}+n\frac{h_1}{n^{k_1}}\right)}\sum_{i=1}^{a}(Z_{i.}-\theta_3)$$
$$= -\frac{\sqrt{\theta_2}h_3}{\left(n^{k_3-\frac{1}{2}}\theta_2+n^{k_3-k_2-\frac{1}{2}}h_2+n^{k_3+\frac{1}{2}-k_1}h_1\right)}\sum_{i=1}^{a}\frac{\sqrt{n}(Z_{i.}-\theta_3)}{\sqrt{\theta_2}}$$
$$\Rightarrow \begin{cases} 0 & (k_3 \le \frac{1}{2},\ k_1 < \frac{1}{2}+k_3) \text{ or } (k_3 > \frac{1}{2}), \\ \frac{\sqrt{\theta_2}h_3}{h_1}g_1(\boldsymbol{T}_n) & k_3 < \frac{1}{2},\ k_1 = \frac{1}{2}+k_3, \\ +\infty \text{ or } -\infty & k_3 < \frac{1}{2},\ k_1 > \frac{1}{2}+k_3, \\ \frac{h_3\sqrt{\theta_2}}{\theta_2+h_1}g_1(\boldsymbol{T}_n) & k_3 = \frac{1}{2}, k_1 = 1, \\ \frac{h_3}{\sqrt{\theta_2}}g_1(\boldsymbol{T}_n) & k_3 = \frac{1}{2},\ k_1 > 1, \end{cases} \tag{6.5}$$

as $n \to \infty$,

(ii)

$$L_2 = -\frac{ah_3^2}{2n^{2k_3-1}\left(\theta_2+\frac{h_2}{n^{k_2}}+n\frac{h_1}{n^{k_1}}\right)}$$
$$= -\frac{ah_3^2}{2\left(n^{2k_3-1}\theta_2+h_2n^{2k_3-1-k_2}+n^{2k_3-k_1}h_1\right)}$$
$$\Rightarrow \begin{cases} 0 & (k_3 \le \frac{1}{2},\ k_1 < 2k_3) \text{ or } (k_3 > \frac{1}{2}), \\ -\frac{ah_3^2}{2h_1} & k_3 < \frac{1}{2},\ k_1 = 2k_3, \\ -\infty & k_3 < \frac{1}{2},\ k_1 > 2k_3, \\ -\frac{ah_3^2}{2(\theta_2+h_1)} & k_3 = \frac{1}{2},\ k_1 = 1, \\ -\frac{ah_3^2}{2\theta_2} & k_3 = \frac{1}{2},\ k_1 > 1, \end{cases}$$

as $n \to \infty$,

(iii)

$$L_2 = -\frac{h_2 a}{2\theta_2 n^{k_2-1}}\left(1-\frac{1}{n}\right) + \frac{ah_2\left(1-\frac{1}{n}\right)}{2(n^{k_2-1}\theta_2+\frac{h_2}{n})}$$

$$+\frac{h_2\sqrt{2a\left(1-\frac{1}{n}\right)}}{2\left(n^{k_2-\frac{1}{2}}\theta_2+\frac{h_2}{\sqrt{n}}\right)}\left\{\frac{\frac{1}{\theta_2}\sum_{i=1}^{a}\sum_{t=1}^{n}(Z_{ij}-Z_{i\cdot})^2-a(n-1)}{\sqrt{2a(n-1)}}\right\}$$

$$+O(n^{1-2k_2})$$

$$\Rightarrow\begin{cases}\text{indeterminate form} & 0<k_2<1,\\ 0 & k_2\geq 1,\end{cases}$$

as $n\to\infty$,

(iv)

$$\begin{aligned}
L_4 &= -\frac{a}{2}\log\frac{\theta_2^{(n)}+n\theta_1^{(n)}}{\theta_2}+\frac{n(\theta_2^{(n)}-\theta_2)+n^2\theta_1^{(n)}}{2\theta_2\left(\theta_2^{(n)}+n\theta_1^{(n)}\right)}\sum_{i=1}^{a}(Z_{i.}-\theta_3)^2\\
&= -\frac{a}{2}\log\left(1+\frac{h_2}{n^{k_2}\theta_2}+\frac{h_1}{n^{k_1-1}\theta_2}\right)\\
&\quad+\left\{\frac{h_2}{2\left(n^{k_2}\theta_2+h_2+n^{k_2-k_1+1}h_1\right)}+\frac{h_1}{2\left(n^{k_1-1}\theta_2+n^{k_1-k_2-1}h_2+h_1\right)}\right\}\\
&\quad\times\sum_{i=1}^{a}\left(\frac{\sqrt{n}(Z_{i.}-\theta_3)}{\sqrt{\theta_2}}\right)^2\\
&\Rightarrow\begin{cases}-\infty & 0<k_1<1,\\ \frac{h_1}{2(\theta_2+h_1)}g_2(\boldsymbol{T}_n)-\frac{a}{2}\log(1+\frac{h_1}{\theta_2}) & k_1=1,\\ 0 & k_1>1,\end{cases}
\end{aligned}$$

as $n\to\infty$.

Here, we employed the Taylor expansion in (ii) and exploited the fact that, under the null H_{61}, $\sqrt{n}(Z_{i.}-\theta_3)/\sqrt{\theta_2}$ follows the standard normal distribution in (i), $\sum_{i=1}^{a}\sqrt{n}(Z_{i.}-\theta_3)^2/\theta_2$ follows the chi-square distribution with a degrees of freedom in (iv), and $\sum_{i=1}^{a}\sum_{t=1}^{n}(Z_{ij}-Z_{i\cdot})^2/\theta_2$ follows the chi-square distribution with $a(n-1)$ degrees of freedom (iii).

Remark 6.1 The log-likelihood ratio process $\Lambda(\boldsymbol{\theta}_{0,61},\boldsymbol{\theta}_{n,61})$ tends to an indeterminate form as $n\to\infty$ on the set $\{k_2;\,0<k_2<1\}$ of related contiguous orders. On the other hand, $\Lambda(\boldsymbol{\theta}_{0,61},\boldsymbol{\theta}_{n,61})$ tends to $-\infty$ as $n\to\infty$ on the set $\{(k_1,k_2,k_3);\,k_2\geq 1,\ k_1<1,k_1\leq k_3+\frac{1}{2}\}$. Due to the term L_1 defined in (6.5), $\Lambda(\boldsymbol{\theta}_{0,61},\boldsymbol{\theta}_{n,61})$ tends to $-\infty$ or to an indeterminate form as $n\to\infty$ on the set $\{(k_1,k_2,k_3);\,k_2\geq 1,k_3<\frac{1}{2},\ k_1>k_3+\frac{1}{2}\}$.

Remark 6.2 Theorem 6.1 tells us that the asymptotic behavior of the likelihood ratio process for the random effect model is non-standard. We wish that the random effect model has the LAN structure in order to construct the optimal test based on the

LAN, but the random effect model does not hold the LAN property. The asymptotic distribution of $\Lambda(\boldsymbol{\theta}_{0,61}, \boldsymbol{\theta}_{n,61})$ includes $g_1\left((x_1, \ldots, x_a)^\top\right)$ and $g_2\left((x_1, \ldots, x_a)^\top\right)$, which tend to the centered normal distribution with variance a and the chi-square distribution with a degrees of freedom, respectively, on the set $\{k_2 \geq 1,\ k_3 = \frac{1}{2},\ k_1 = 1\}$. The Fisher information matrix is singular.

From Theorem 6.1 and Remark 6.2, we cannot use the LAN framework for our setting, and thus, a different approach is needed. Fortunately, the Neyman–Pearson lemma leads to an asymptotically most powerful test (see Definition 6.1). For simplicity, we shall consider the simpler contiguous hypothesis:

$$\mathrm{H}_{62} : \boldsymbol{\theta} = \boldsymbol{\theta}_{0,62} := \begin{pmatrix} 0 \\ \theta_2 \\ \theta_3 \end{pmatrix} \quad \text{vs} \quad \mathrm{K}_{62}^{(n)} : \boldsymbol{\theta} = \boldsymbol{\theta}_{n,62} := \begin{pmatrix} \frac{h_1}{n^{k_1}} \\ \theta_2 \\ \theta_3 \end{pmatrix},$$

where $\theta_2 > 0$ and $h_1 > 0$.

Then, the simpler result corresponding to Theorem 6.1 holds.

Theorem 6.2 *Under the null hypothesis* H_{62}, $\Lambda(\boldsymbol{\theta}_{0,62}, \boldsymbol{\theta}_{n,62})$ *has the following the asymptotic expansion:*

$$\Lambda(\boldsymbol{\theta}_{0,62}, \boldsymbol{\theta}_{n,62}) = \begin{cases} -\frac{a}{2}\log(1 + \frac{h_1}{\theta_2}) + \frac{h_1}{2(\theta_2+h_1)} g_2(\boldsymbol{T}_{1,\mathrm{GKKT},n}) + o_p(1) & k_1 = 1, \\ o_p(1) & k_1 > 1. \end{cases}$$

Proof ***(Theorems*** *6.2)*

Let the elements of $\boldsymbol{\theta}_{n,62}$ denote $\left(\theta_1^{(n)}\ \theta_2\ \theta_3\right)^\top$. Then, from the definition of $\Lambda(\boldsymbol{\theta}_{0,62}, \boldsymbol{\theta}_{n,62})$,

$$\begin{aligned} \Lambda(\boldsymbol{\theta}_{0,62}, \boldsymbol{\theta}_{n,62}) &= -\frac{a}{2}\log\frac{\theta_2 + n\theta_1^{(n)}}{\theta_2} + \frac{n^2\theta_1^{(n)}}{2\theta_2(\theta_2 + n\theta_1^{(n)})}\sum_{i=1}^{a}(Z_{i\cdot} - \theta_3)^2 \\ &= -\frac{a}{2}\log\left(1 + \frac{h_1}{\theta_2 n^{k_1-1}}\right) + \frac{h_1}{2(n^{k_1-1}\theta_2 + h_1)}\frac{n}{\theta_2}\sum_{i=1}^{a}(Z_{i\cdot} - \theta_3)^2, \end{aligned}$$

which gives the conclusion. □

Remark 6.3 The log-likelihood ratio process $\Lambda(\boldsymbol{\theta}_0, \boldsymbol{\theta}_n)$ tends to $-\infty$ as $n \to \infty$ on the set $\{k_1; k_1 < 1\}$.

The asymptotic distribution of $\Lambda(\boldsymbol{\theta}_{0,62}, \boldsymbol{\theta}_{n,62})$ under the alternative $\mathrm{K}_{62}^{(n)}$ can be derived as follows.

Theorem 6.3 *Under the null hypothesis* $\mathrm{K}_{62}^{(n)}$, *the following the asymptotic expansion holds:*

$$\Lambda(\boldsymbol{\theta}_{0,62}, \boldsymbol{\theta}_{n,62}) = \begin{cases} -\frac{a}{2}\log(1+\frac{h_1}{\theta_2}) + \frac{h_1}{2\theta_2} g_2(\boldsymbol{T}_{2,\mathrm{GKKT},n}) + o_p(1) & k_1 = 1, \\ o_p(1) & k_1 > 1, \end{cases}$$

as $n \to \infty$, where

$$\boldsymbol{T}_{2,\mathrm{GKKT},n} := \left(\frac{\sqrt{n}(Z_{1\cdot} - \theta_3)}{\sqrt{n\theta_1^{(n)} + \theta_2}}, \ldots, \frac{\sqrt{n}(Z_{a\cdot} - \theta_3)}{\sqrt{n\theta_1^{(n)} + \theta_2}} \right)^{\top}, \tag{6.6}$$

which follows, under $\mathrm{H}_{61}^{(n)}$, the a-dimensional standard normal distribution.

Proof (Theorems 6.3) Let the elements of $\boldsymbol{\theta}_{n,62}$ denote $\left(\theta_1^{(n)}\ \theta_2\ \theta_3\right)^{\top}$. From the definition of $\Lambda(\boldsymbol{\theta}_{0,62}, \boldsymbol{\theta}_{n,62})$,

$$\begin{aligned} \Lambda(\boldsymbol{\theta}_{0,62}, \boldsymbol{\theta}_{n,62}) &= -\frac{a}{2}\log\frac{\theta_2 + n\theta_1^{(n)}}{\theta_2} + \frac{n^2\theta_1^{(n)}}{2\theta_2(\theta_2 + n\theta_1^{(n)})}\sum_{i=1}^{a}(Z_{i\cdot} - \theta_3)^2 \\ &= -\frac{a}{2}\log\left(1 + \frac{h_1}{\theta_2 n^{k_1-1}}\right) + \frac{h_1}{2\theta_2 n^{k_1-1}}\frac{n}{\theta_2 + n\theta_1^{(n)}}\sum_{i=1}^{a}(Z_{i\cdot} - \theta_3)^2, \end{aligned}$$

which gives the conclusion. □

Remark 6.4 The log-likelihood ratio process $\Lambda(\boldsymbol{\theta}_{0,62}, \boldsymbol{\theta}_{n,62})$ tends to an indeterminate form as $n \to \infty$ on the set $\{k_1; k_1 < 1\}$.

To state the main result, let us define the asymptotically most powerful test at the asymptotic level α Lehmann and Romano (2006, Definition 13.3.1, p. 541).

Definition 6.1 For the simple hypothesis $\vartheta = \vartheta_0$ against $\vartheta = \vartheta_n := \vartheta_0 + hn^{-k}$ for some $k > 0$ and $h > 0$, a sequence of tests $\{\phi_n\}$ is asymptotically most powerful at the asymptotic level α if

$$\limsup_n \mathrm{E}_{\vartheta_0}(\phi_n) \leq \alpha,$$

and, for any test $\{\psi_n\}$ such that $\limsup_n \mathrm{E}_{\vartheta_0}(\psi_n) \leq \alpha$,

$$\limsup_n \left(\mathrm{E}_{\vartheta_n}(\phi_n) - \mathrm{E}_{\vartheta_n}(\psi_n)\right) \geq 0.$$

By recalling the Neyman–Pearson lemma, we define the test function, for α_n such that $\alpha_n \to \alpha$ as $n \to \infty$, as

$$\phi_{\mathrm{GKKT},n} := \begin{cases} 1 & \Lambda(\boldsymbol{\theta}_{0,62}, \boldsymbol{\theta}_{n,62}) > c_n, \\ \gamma_n & \Lambda(\boldsymbol{\theta}_{0,62}, \boldsymbol{\theta}_{n,62}) = c_n, \\ 0 & \Lambda(\boldsymbol{\theta}_{0,62}, \boldsymbol{\theta}_{n,62}) < c_n, \end{cases} \tag{6.7}$$

where the critical value c_n and the constant γ_n are determined by $\mathrm{E}_{\theta_0}(\phi_{\mathrm{GKKT},n}) = \alpha_n$.

Then, we can show that the test defined in (6.7) is asymptotically most powerful.

Theorem 6.4 *Fix $k_1 = 1$. Then,*

(i) the critical value c_n converges to

$$c := -\log\frac{a}{2}\log\left(1+\frac{h_1}{\theta_2}\right)+\frac{h_1}{2(\theta_2+h_1)}\chi_a^2[1-\alpha] \quad \text{as } n\to\infty,$$

where $\chi_a^2[1-\alpha]$ denotes the upper α quantile of chi-square with a degrees of freedom,

(ii) the asymptotic power of the test under the local alternative $\mathrm{K}_3^{(n)}$ is given by

$$\lim_{n\to\infty} P\left(\Lambda(\boldsymbol{\theta}_{0,62},\boldsymbol{\theta}_{n,62})\geq c_n\right) = P\left(\mathfrak{X}\geq\frac{\theta_2}{(\theta_2+h_1)}\chi_a^2[1-\alpha]\right),$$

where $\mathfrak{X}$ follows the chi-square distribution with a degrees of freedom, and

(iii) the test $\{\phi_{\mathrm{GKKT},n}\}$ is asymptotically most powerful at asymptotic level α.

***Proof (Theorems** 6.4)* Denote the asymptotic distributions of the log-likelihood ratio process $\Lambda(\boldsymbol{\theta}_{0,62},\boldsymbol{\theta}_{n,62})$ under the null and alternative by $\mathfrak{L}_H$ and $\mathfrak{L}_K$.

(i) From Lemma 2.11 of Van der Vaart (2000), Theorem 6.2 yields, under the null H_{62},

$$\sup_{x\in\mathbb{R}}\left|\mathrm{P}\left(\Lambda(\boldsymbol{\theta}_{0,62},\boldsymbol{\theta}_{n,62})<x\right)-\mathrm{P}(\mathfrak{L}_H<x)\right|\to 0 \quad \text{as } n\to\infty,$$

and, thus,

$$\alpha_n = P(\Lambda(\boldsymbol{\theta}_{0,62},\boldsymbol{\theta}_{n,62})\geq c_n)\to P(\mathfrak{L}_H\geq c) \quad \text{as } n\to\infty,$$

where c is a constant satisfying $P(\mathfrak{L}_H\geq c)=\alpha$. This implies

$$c_n\to c := -\log\frac{a}{2}\log\left(1+\frac{h_1}{\theta_2}\right)+\frac{h_1}{2(\theta_2+h_1)}\chi_a^2[1-\alpha] \quad \text{as } n\to\infty.$$

(ii) Apply Theorems 6.3 and 6.4 (i) to deduce

$$\Lambda(\boldsymbol{\theta}_{0,62},\boldsymbol{\theta}_{n,62})-c_n\Rightarrow\mathfrak{L}_K-c \quad \text{under } \mathrm{K}_{62}^{(n)},$$

and, thus, under $\mathrm{K}_{62}^{(n)}$, we have

$$\begin{aligned}\lim_{n\to\infty}\mathrm{P}(\Lambda(\boldsymbol{\theta}_{0,62},\boldsymbol{\theta}_{n,62})\geq c_n) &= \mathrm{P}(\mathfrak{L}_K\geq c)\\ &= \mathrm{P}\left(\mathfrak{X}\geq\frac{\theta_2}{(\theta_2+h_1)}\chi_a^2[1-\alpha]\right).\end{aligned}$$

(iii) Applying the Neyman–Pearson lemma, we obtain, for any $n \in \mathbb{N}$ and any test $\{\psi_n\}$ such that $\limsup_n \mathrm{E}_{\theta_n}(\psi_n) \le \alpha$, that $\mathrm{E}_{\theta_n}(\phi_{\mathrm{GKKT},n}) - \mathrm{E}_{\theta_n}(\psi_n) \ge 0$.

□

6.3 The Hypothesis That the Variance of Random Effects is not Equal to Zero Under the Null

In this section, we consider the contiguous hypothesis that the variance of random effects belongs to the interior of the parameter space under the null, which can be written as

$$\mathrm{H}_{63} : \boldsymbol{\theta} = \boldsymbol{\theta}_{0,63} := \begin{pmatrix} \theta_1 \\ \theta_2 \\ \theta_3 \end{pmatrix}, \quad \mathrm{K}_{63}^{(n)} : \boldsymbol{\theta} = \boldsymbol{\theta}_{n,63} := \begin{pmatrix} \theta_1 + \frac{h_1}{n^{k_1}} \\ \theta_2 + \frac{h_2}{n^{k_2}} \\ \theta_3 + \frac{h_3}{n^{k_3}} \end{pmatrix},$$

where $\theta_1 > 0, \theta_2 > 0, h_1 > 0, h_1 > -n^{k_1}\theta_1$, and $h_2 > -n^{k_2}\theta_2$.

Theorem 6.5 *Under the null hypothesis* H_{63}*, for all* $k_1 \ge 1$*,* $k_2 > 0$*, and* $k_3 > 0$*, the log-likelihood ratio* $\Lambda(\boldsymbol{\theta}_{0,63}, \boldsymbol{\theta}_{n,63})$ *degenerates to zero, that is,* $\Lambda(\boldsymbol{\theta}_{0,63}, \boldsymbol{\theta}_{n,63})$ *converges in probability to zero as* $n \to \infty$.

Proof *(Proof of Theorem* 6.5*)*

Let the elements of $\boldsymbol{\theta}_{n,63}$ denote $\left(\theta_1^{(n)}\ \theta_2^{(n)}\ \theta_3^{(n)}\right)^\top$. A lengthy but straightforward computation gives that

$$\begin{aligned}
&\Lambda(\boldsymbol{\theta}_{0,63}, \boldsymbol{\theta}_{n,63}) \\
&= -\frac{a}{2}(n-1)\log\frac{\theta_2^{(n)}}{\theta_2} - \frac{a}{2}\log\frac{\theta_2^{(n)} + n\theta_1^{(n)}}{\theta_2 + n\theta_1} + \frac{\theta_2^{(n)} - \theta_2}{2\theta_2^{(n)}}\left(\frac{1}{\theta_2}\sum_{i=1}^{a}\sum_{t=1}^{n}(Z_{it} - Z_{i\cdot})^2\right) \\
&\quad + \frac{(\theta_2^{(n)} - \theta_2) + n(\theta_1^{(n)} - \theta_1)}{2(\theta_2^{(n)} + n\theta_1^{(n)})}\left(\frac{n}{\theta_2 + n\theta_1}\sum_{i=1}^{a}(Z_{i\cdot} - \theta_3)^2\right) \\
&\quad + \frac{\sqrt{n\theta_2 + n^2\theta_1}(\theta_3^{(n)} - \theta_3)}{\theta_2^{(n)} + n\theta_1^{(n)}}\left(\sum_{i=1}^{a}\frac{\sqrt{n}\,(Z_{i\cdot} - \theta_3)}{\sqrt{\theta_2 + n\theta_1}}\right) - \frac{an(\theta_3^{(n)} - \theta_3)^2}{2(\theta_2^{(n)} + n\theta_1^{(n)})} \\
&= -\frac{a}{2}(n-1)\log\left(1 + \frac{h_2}{n^{k_2}\theta_2}\right) - \frac{a}{2}\log\left(1 + \frac{h_2}{n^{k_2}\theta_2 + n^{k_2+1}\theta_1} + \frac{h_1}{n^{k_1-1}\theta_2 + n^{k_1}\theta_1}\right) \\
&\quad + \frac{h_2\sqrt{2a\left(1 - \frac{1}{n}\right)}}{2\left(n^{k_2-\frac{1}{2}}\theta_2 + \frac{h_2}{\sqrt{n}}\right)}\left(\frac{\frac{1}{\theta_2}\sum_{i=1}^{a}\sum_{t=1}^{n}(Z_{ij} - Z_{i\cdot})^2 - a(n-1)}{\sqrt{2a(n-1)}}\right) + \frac{ah_2(1 - \frac{1}{n})}{2(n^{k_2-1}\theta_2 + \frac{h_2}{n})} \\
&\quad + \left\{\frac{h_2}{2(n^{k_2}\theta_2 + h_2 + n^{k_2+1}\theta_1 + n^{k_2-k_1+1}h_1)}\right.
\end{aligned}$$

$$+\frac{h_1}{2(n^{k_1-1}\theta_2+n^{k_1-k_2-1}h_2+n^{k_1}\theta_1+h_1)}\Bigg\}\sum_{i=1}^{a}\left(\frac{Z_{i.}-\sqrt{n}\theta_3}{\sqrt{\theta_2+n\theta_1}}\right)^2$$

$$+\frac{h_3\sqrt{\frac{\theta_2}{n}+\theta_1}}{n^{k_3-1}\theta_2+n^{k_3-k_2-1}h_2+n^{k_3}\theta_1+n^{k_3-k_1}h_1}\sum_{i=1}^{a}\frac{Z_{i.}-\sqrt{n}\theta_3}{\theta_2+n\theta_1}$$

$$-\frac{ah_3^2}{2(n^{2k_3-1}\theta_2+n^{2k_3-k_2-1}h_2+n^{2k_3}\theta_1+n^{2k_3-k_1}h_1)},$$

which, in conjunction with the fact that $\sum_{i=1}^{a}(\sqrt{n}(Z_{i\cdot}-\theta_3)/\sqrt{\theta_2+n\theta_1})^2$ follows the chi-square distribution with a degrees of freedom under the null hypothesis H_{63}, tends in probability to zero as $n\to\infty$ on the set $\{k_1;\, k_1\geq 1\}$. □

Remark 6.5 For $\{k_1;\, k_1<1\}$, $\Lambda(\boldsymbol{\theta}_{0,63},\boldsymbol{\theta}_{n,63})$ tends to an indeterminate form as $n\to\infty$.

Remark 6.6 Theorem 6.5 shows that the asymptotic behavior of the log-likelihood ratio process $\Lambda(\boldsymbol{\theta}_{0,63},\boldsymbol{\theta}_{n,63})$ is non-standard even if the variance of the random effects belongs to the interior of the parameter space under the null.

Remark 6.7 For the contiguous hypothesis

$$\mathrm{H}_{64}:\boldsymbol{\theta}=\boldsymbol{\theta}_0:=\begin{pmatrix}\theta_1\\ \theta_2\\ \theta_3\end{pmatrix},\quad \mathrm{K}_{64}^{(n)}:\boldsymbol{\theta}=\boldsymbol{\theta}_n:=\begin{pmatrix}\theta_1+\frac{h_1}{n^{k_1}}\\ \theta_2\\ \theta_3\end{pmatrix},$$

where $\theta_1>0$, $\theta_2>0$, $h_1>0$, and $h_1>-n^{k_1}\theta_1$, we can show the same result as Theorem 6.5, that is, under the null hypothesis H_{64}, $\Lambda(\boldsymbol{\theta}_{0,64},\boldsymbol{\theta}_{n,64})$ converges in probability to zero as $n\to\infty$.

***Proof (Remark** 6.7)* Let $\boldsymbol{\theta}_{n,64}=\left(\theta_1^{(n)}\ \theta_2\ \theta_3\right)^\top$. A straightforward calculation yields

$$\begin{aligned}
&\Lambda(\boldsymbol{\theta}_{0,64},\boldsymbol{\theta}_{n,64})\\
&=-\frac{a}{2}\log\frac{\theta_2+n\theta_1^{(n)}}{\theta_2+n\theta_1}+\frac{n(\theta_1^{(n)}-\theta_1)}{2(\theta_2+n\theta_1^{(n)})}\left(\frac{n}{\theta_2+n\theta_1}\sum_{i=1}^{a}(Z_{i\cdot}-\theta_3)^2\right)\\
&=-\frac{a}{2}\log\left(1+\frac{h_1}{n^{k_1-1}\theta_2+n^{k_1}\theta_1}\right)\\
&\quad+\frac{h_1}{2(n^{k_1-1}\theta_2+n^{k_1}\theta_1+h_1)}\left(\frac{n}{\theta_2+n\theta_1}\sum_{i=1}^{a}(Z_{i.}-\theta_3)^2\right),
\end{aligned}$$

which tends to 0 as $n\to\infty$.

□

References

Goto, Y., Kaneko, T., Kojima, S., & Taniguchi, M. (2022). Likelihood ratio processes under non-standard settings. *Theory of Probability and Its Applications, 67,* 246–260.

Hallin, M., Hlubinká, D., & Hudecova, S. (2021). Efficient fully distribution-free center-outward rank tests for multiple-output regression and MANOVA. *Journal of the American Statistical Association, na*, 1–43.

Lehmann, E. L., & Romano, J. P. (2006). *Testing statistical hypotheses*. Springer.

Searle, S. R., Casella, G., & McCulloch, C. E. (1992). *Variance components*. New York: Wiley.

Van der Vaart, A. W. (2000). *Asymptotic statistics 3*. Cambridge University Press.

Chapter 7
Numerical Analysis

In this chapter, we illustrate some numerical studies investigating the finite sample performance of the tests for one-way and two-way models presented in the previous chapters. First, we demonstrate the performance of the Lawley–Hotelling test statistic (LH), the likelihood ratio test statistic (LR), and the Bartlett–Nanda–Pillai test statistic (BNP), defined in (2.5)–(2.7), respectively, for Chapter 2. Second, we investigate the finite sample performance of the tests based on (4.6) and (4.7) in Chapter 4. Note that the test statistic defined in (4.6) is almost the same as that defined in (2.16). Third, we illustrate the performance of the test based on (5.5) and (5.8) in Chapter 5.

7.1 One-Way Effect Model with Independent Groups

We consider the one-way fixed effect model (see also (2.1)):

$$\boldsymbol{Z}_{it} = \boldsymbol{\mu} + \boldsymbol{\alpha}_i + \boldsymbol{\epsilon}_{it} \qquad t = 1, \ldots, n_i, \qquad i = 1, \ldots, a,$$

where $\boldsymbol{Z}_{it}$ is a p-dimensional random vector, $\boldsymbol{\mu} = (\mu_1, \ldots, \mu_p)^T$, $\boldsymbol{\alpha}_i = (\alpha_{i,1}, \ldots, \alpha_{i,p})^T$, and $\boldsymbol{\epsilon}_{it}$ is the disturbance term.

Recall that the hypothesis for the treatment effects given in (2.3) is

$$H_{21} : \boldsymbol{\alpha}_1 = \boldsymbol{\alpha}_2 = \ldots = \boldsymbol{\alpha}_a = \boldsymbol{0} \text{ vs } K_{21} : \boldsymbol{\alpha}_i \neq \boldsymbol{0} \text{ for some } i.$$

The null hypothesis H_{21} implies that all the effects are zero.

For $\boldsymbol{\epsilon}_{it} = \left(\epsilon_{it}^{(1)}, \ldots, \epsilon_{it}^{(p)}\right)^\top$, the DCC-GARCH(1, 1) is a typical example of an uncorrelated process (see Engle, 2002). We assume the DCC-GARCH(1,1) model for $\boldsymbol{\epsilon}_{it}$, that is, $\boldsymbol{\epsilon}_{it}$ follows p-dimensional centered normal distribution with variance $\boldsymbol{H}_{i,t}$, where

$$\boldsymbol{H}_{it} = \boldsymbol{D}_{it}\boldsymbol{R}_{it}\boldsymbol{D}_{it}, \qquad \boldsymbol{D}_{it} = \text{diag}\left[\sqrt{\sigma_{it}^{(1)}}, \ldots, \sqrt{\sigma_{it}^{(p)}}\right],$$

Y. Goto et al., *ANOVA with Dependent Errors*, JSS Research Series in Statistics,
https://doi.org/10.1007/978-981-99-4172-8_7

$$\sigma_{it}^{(j)} = c_j + a_j \left\{\epsilon_{i(t-1)^{(j)}}\right\}^2 + b_j \sigma_{i(t-1)}^{(j)}, \tag{7.1}$$

$$\boldsymbol{R}_{it} = \left(diag\left[\boldsymbol{Q}_{it}\right]\right)^{-1/2} \boldsymbol{Q}_{it} \left(diag\left[\boldsymbol{Q}_{it}\right]\right)^{-1/2},$$

$$\boldsymbol{Q}_{it} = (1-\alpha-\beta)\ \tilde{\boldsymbol{Q}} + \alpha\ \tilde{\boldsymbol{\epsilon}}_{i(t-1)}\tilde{\boldsymbol{\epsilon}}_{i(t-1)}^{T} + \beta\ \boldsymbol{Q}_{it-1}, \tag{7.2}$$

$$\tilde{\boldsymbol{\epsilon}}_{it} = \left(\tilde{\epsilon}_{it}^{(1)}, \ldots, \tilde{\epsilon}_{it}^{(p)}\right)^T, \text{ and } \tilde{\epsilon}_{it}^{(j)} = \frac{\epsilon_{it}^{(j)}}{\sqrt{\sigma_{it}^{(j)}}}, \quad j = 1, \ldots, p.$$

7.1.1 The Empirical Size

This subsection focuses on the tests under the null hypothesis. We carry out two experiments to study the finite sample performance of the LH, the LR, and the BNP test under the null (2.3). In the first experiment, we investigate the performance of three tests with different p and sample sizes under fixed $a = 3$. In specific, we set the dimension p of $\boldsymbol{Z}_{it}$ to vary between 2 and 5 and consider $a = 3$ groups with three samples: $n_1 = 1000$, $n_2 = 1500$, and $n_3 = 2000$. The parameters of Eq. (7.1) are given in Appendix at the end of this chapter along with the unconditional correlation matrix $\tilde{\boldsymbol{Q}}$. In (7.2), we set $\alpha = 0.05$ and $\beta = 0.85$.

The empirical size from 10,000 replications is presented in Table 7.1 for the LH, LR, and BNP tests. It can be seen that, under the null hypothesis, regardless of different p values and various nominal levels, the actual level is always remarkably close to the nominal level, and the performance of the three tests is very similar. For example, for the LH test, given the nominal level 0.1, the simulated significance level varies from 0.102 to 0.107 with p changing from 2 to 5, which are all quite close to 0.1. In summary, all the classical tests are effective and varying the dimension of $\boldsymbol{Z}_{it}$ does not affect the accuracy of the three tests.

Table 7.1 Empirical size of the LH, LR, and BNP tests when p varies ($a = 3$)

	Nominal	Dimension of $\boldsymbol{Z}_{it}$			
	level	$p = 2$	$p = 3$	$p = 4$	$p = 5$
LH	0.10	0.102	0.104	0.102	0.107
	0.05	0.053	0.054	0.054	0.054
	0.01	0.009	0.012	0.010	0.011
LR	0.10	0.102	0.104	0.102	0.105
	0.05	0.053	0.054	0.053	0.054
	0.01	0.009	0.012	0.010	0.011
BNP	0.10	0.102	0.103	0.102	0.105
	0.05	0.053	0.054	0.053	0.053
	0.01	0.009	0.012	0.010	0.010

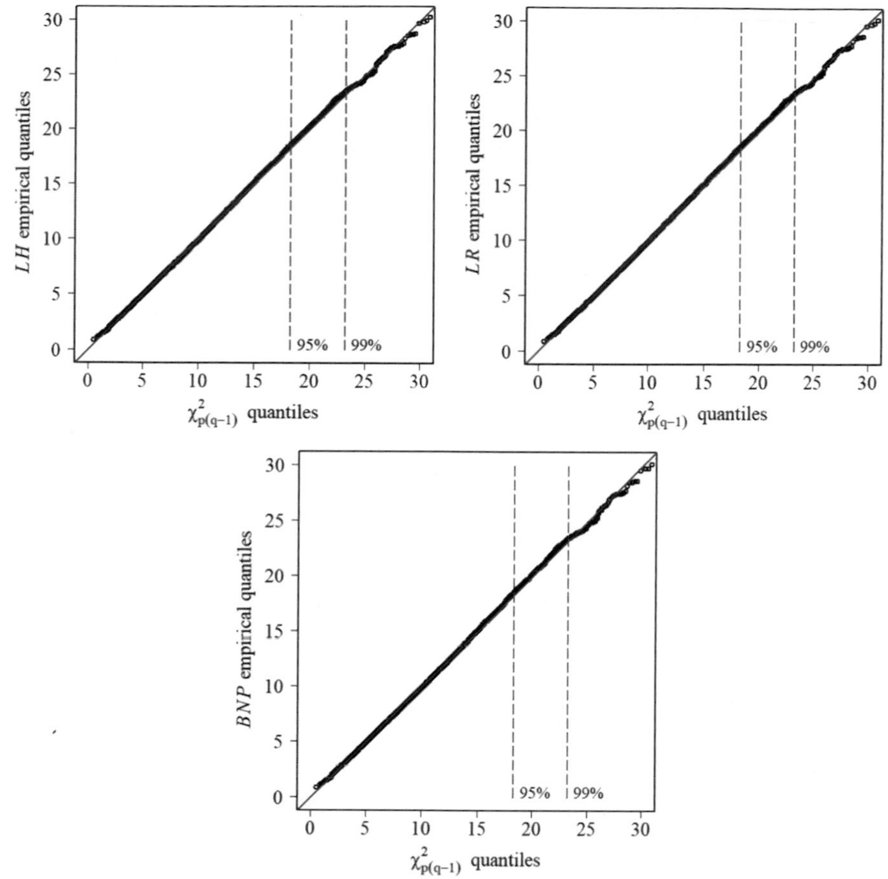

Fig. 7.1 QQ-plot of the LH (left upper panel), LR (right upper panel), and BNP (lower panel) statistics versus the chi-square distribution with $p(a-1)$ degrees of freedom under H_0 when $p = 5$ and $a = 3$

The QQ-plots of the LH, LR, and BNP test statistics versus the chi-square distribution with $p(a-1)$ degrees of freedom, under H_0, when $p = 5$ and $a = 3$ are in Fig. 7.1. It is evident that the distribution of the three statistics is approximated extremely well by the chi-square distribution with $p(a-1)$ degrees of freedom even far away in the tails.

In the second simulation experiment, we fix $p = 2$ while the number a of groups varies, i.e., $a = 3, 4, 5, 7, 10$, and the sample sizes are set as follows:

- $n_1 = 1000, n_2 = 1500, n_3 = 2000$ for $a = 3$
- $n_1 = 1000, n_2 = 1300, n_3 = 1700, n_4 = 2000$ for $a = 4$
- $n_1 = 1000, n_2 = 1250, n_3 = 1500, n_4 = 1750, n_5 = 2000$ for $a = 5$
- $n_1 = 1000,\ n_2 = 1250,\ n_3 = 1500,\ n_4 = 1750,\ n_5 = 2000,\ n_6 = 2250,\ n_7 = 2500$ for $a = 7$

Table 7.2 Empirical size of the LH, LR, and BNP tests when a varies ($p = 2$)

	Nominal	Number of groups				
	level	$a = 3$	$a = 4$	$a = 5$	$a = 7$	$a = 10$
LH	0.10	0.102	0.098	0.100	0.096	0.102
	0.05	0.053	0.048	0.049	0.047	0.053
	0.01	0.009	0.010	0.011	0.008	0.009
LR	0.10	0.102	0.098	0.100	0.096	0.102
	0.05	0.053	0.048	0.048	0.047	0.053
	0.01	0.009	0.010	0.011	0.008	0.009
BNP	0.1	0.102	0.097	0.100	0.095	0.102
	0.05	0.053	0.048	0.048	0.047	0.053
	0.01	0.009	0.010	0.011	0.008	0.009

- $n_1 = 1000,\ n_2 = 1200,\ n_3 = 1400,\ n_4 = 1600,\ n_5 = 1800,\ n_6 = 2000,\ n_7 = 2200, n_8 = 2400, n_9 = 2600, n_{10} = 2800$ for $a = 10$.

Table 7.2 displays the average simulated level with various q of the LH, LR, and BNP tests over 10,000 replications. It shows that whatever the number of groups, the actual level of the three tests is again very close to the nominal level under the null hypothesis. In other words, all the classical tests are robust to the number of groups.

The QQ-plots of the LH, LR, and BNP statistics versus the chi-square distribution with $p(a-1)$ degrees of freedom, under H_0, when $p = 2$ and $a = 10$ are displayed in Fig. 7.2. It can be seen that the distribution of the three statistics is impressively close to the chi-square distribution with $p(a-1)$ degrees of freedom even for a number of groups as large as 10.

7.1.2 The Power

To investigate the power of the tests under the alternative hypothesis, we consider the same setup as 7.1 but we set fixed effects $\alpha_i = \delta_i \mathbf{1}_p$, where $\mathbf{1}_p$ is a p-dimensional unit vector. In case $a = 3$, we fix $\delta_1 = -0.01$, $\delta_2 = 0$, and $\delta_3 = 0.01$. Notice that this is a very small deviation from the null hypothesis. Table 7.3 illustrates the empirical power with nominal levels 0.01, 0.05, and 0.1 when p varies. Unlike the null hypothesis case, it shows that the power of all the three tests is affected by the dimension of $\mathbf{Z}_{it}$, and it shows that the power decreases when p increases for all the three tests. Again the differences in the performances of the three tests are negligible for all the three tests.

Figure 7.3 illustrates the receiver operating characteristic (ROC) curves for the LR, the LH, and the BNP tests when $p = 5$ and $a = 3$. The ROC curve is obtained by plotting the empirical power on y-axis (under the alternative) and the empirical size (under the null) on the x-axis. The closer the ROC curve is to the upper left corner of the graph, the better is the performance of the test.

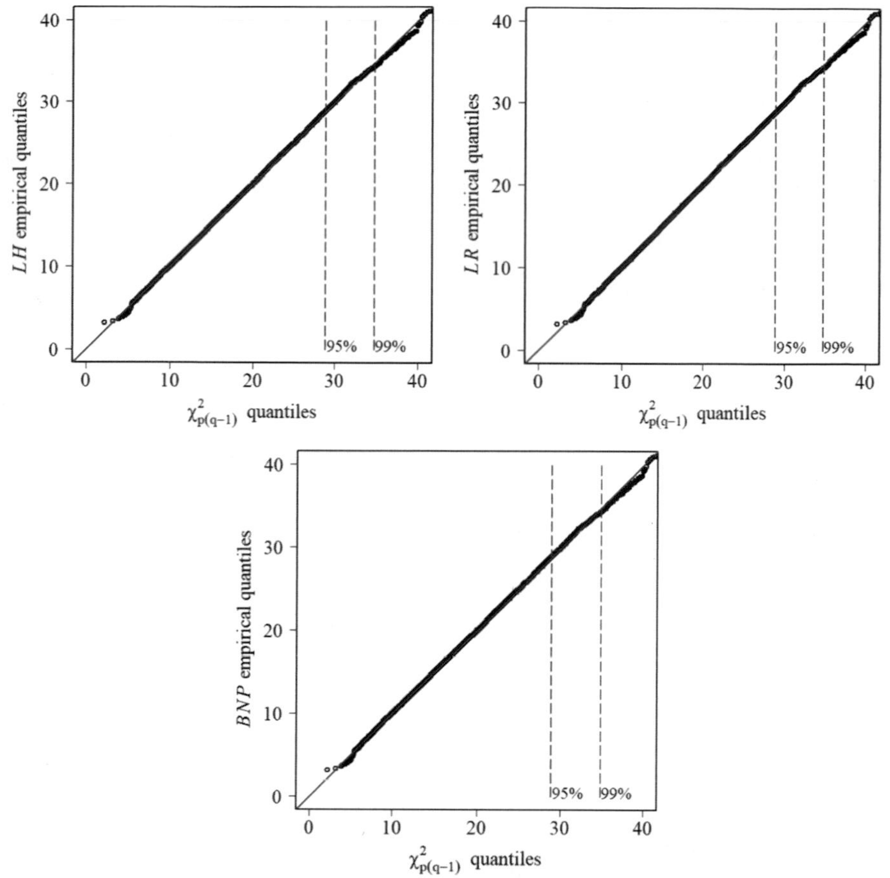

Fig. 7.2 *QQ*-plots of the LH (left upper panel), LR (right upper panel), and BNP (lower panel) statistics versus the chi-square distribution with $p(a-1)$ degrees of freedom under H_0 when $p = 2$ and $a = 10$

Table 7.3 Empirical power of the LR, LH, and BNP tests when p varies ($a = 3$)

		Dimension of $\boldsymbol{Z}_{it}$			
Nominal level	Test	$p = 2$	$p = 3$	$p = 4$	$p = 5$
0.01	LR	0.973	0.651	0.630	0.567
	LH	0.974	0.651	0.632	0.569
	BNP	0.973	0.649	0.629	0.566
0.05	LR	0.948	0.840	0.827	0.784
	LH	0.948	0.841	0.828	0.786
	BNP	0.947	0.839	0.827	0.783
0.1	LR	0.849	0.650	0.630	0.567
	LH	0.849	0.651	0.632	0.569
	BNP	0.848	0.649	0.629	0.566

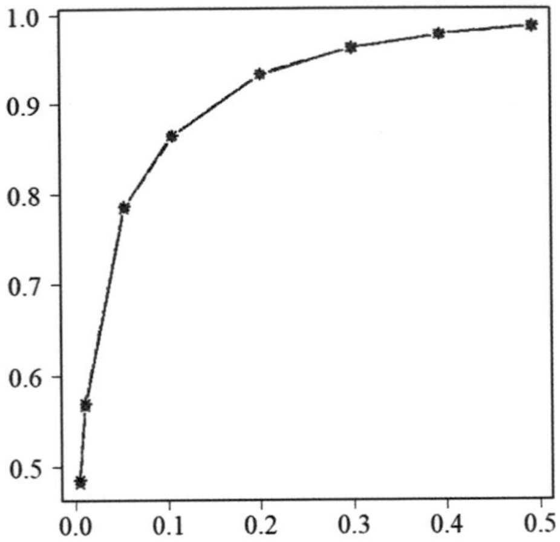

Fig. 7.3 The ROC curve for the LR (solid line with circles), the LH (dashed line with stars), and the BNP (dotted line with crosses) test when $a = 3$ and $p = 5$ (the knots correspond to ξ percentiles of the chi-square distribution with $p(a-1)$, where $\xi = 0.5, 0.6, 0.7, 0.8, 0.9, 0.95, 0.99, 0.995$). The x-axis is the empirical size and the y-axis is the empirical power

It can be seen that the three curves are overlapping which indicates that there is no appreciable difference among the tests.

Next, the power is investigated when a varies with fixed $p = 2$. The values of the δ_i's for $a = 3, 4, 5, 7, 10$ are reported below.

- $a = 3 \Rightarrow \delta_1 = -0.01, \delta_2 = 0, \delta_3 = 0.01$
- $a = 4 \Rightarrow \delta_1 = -0.015, \delta_2 = -0.005, \delta_3 = 0.005, \delta_4 = -0.015$
- $a = 5 \Rightarrow \delta_1 = -0.02, \delta_2 = -0.01, \delta_3 = 0, \delta_4 = 0.01, \delta_5 = 0.02$
- $a = 7 \Rightarrow \delta_1 = -0.03, \delta_2 = -0.03, \delta_3 = -0.01, \delta_4 = 0, \delta_5 = 0.01, \delta_6 = 0.02, \delta_7 = 0.03$
- $a = 10 \Rightarrow \delta_1 = -0.045, \delta_2 = -0.035, \delta_3 = -0.025, \delta_4 = -0.015, \delta_5 = -0.005, \delta_6 = 0.005, \delta_7 = 0.015, \delta_8 = 0.025, \delta_9 = 0.035, \delta_{10} = 0.045$

Table 7.4 presents the simulated power of three tests with nominal levels equal to 0.01, 0.05, and 0.1 when a varies from 3 to 10. It shows that all the three tests

Table 7.4 Empirical power of the LR, LH, and BNP tests when a varies ($p = 2$, nominal level is set as 0.01)

Nominal		Number of groups				
level	Test	$a = 3$	$a = 4$	$a = 5$	$a = 7$	$a = 10$
0.01	LR	0.849	0.999	1.000	1.000	1.000
	LH	0.849	1.000	1.000	1.000	1.000
	BNP	0.848	0.999	1.000	1.000	1.000
0.05	LR	0.948	1.000	1.000	1.000	1.000
	LH	0.948	1.000	1.000	1.000	1.000
	BNP	0.947	1.000	1.000	1.000	1.000
0.1	LR	0.973	1.000	1.000	1.000	1.000
	LH	0.974	1.000	1.000	1.000	1.000
	BNP	0.973	1.000	1.000	1.000	1.000

exhibit strong power under the alternative hypothesis, and increasing the number of groups enhances the power of the tests. For example, when the nominal level equals 0.01 and 0.05 if $a = 3$, the power of all tests is slightly smaller than one, while when a is larger than 3, the power increases to one. In summary, all the three tests show satisfactory performance in terms of simulated level and power, and no substantial differences are observed among them. Changing the dimension p and the number of groups q does not appreciably affect the simulated significance level, while increasing p reduces the power and increasing q enhances the power of these tests.

7.2 One-Way Effect Model with Correlated Groups

In this part, we show the finite sample performance for the test statistic $T_{\mathrm{GALT},n}$ and $T_{\mathrm{ts},n}$ for fixed and random models with time-dependent errors and correlated groups based on Chapter 4.

7.2.1 *Setup*

We generate data from the one-way effect model:

$$\boldsymbol{Z}_{it} = \boldsymbol{\mu} + \boldsymbol{\alpha}_i + \boldsymbol{e}_{it} \quad \text{for } i = 1, \dots, a;\ t = 1, \dots, n_i.$$

We set the dimensions p, q, and the number of groups a as follows: $p = 1, q = 1$, and $a = 3, 6, 9$, respectively. Consider five experiment designs with different setup of group size: (BI) $n_1 = \cdots = n_a = 300$, (BII) $n_1 = \cdots = n_a = 1000$, (BIII) $n_1 = \cdots = n_a = 2000$, (BIV) $n_1 = \cdots = n_a = 3000$, and (Unbalanced) $n_{3k-1} = 500$, and other sample sizes are 2000. That is, (B1) – (BIV) cases are balanced designs with equal group size, and the size varies from small to larger values. (Unbalanced) is the case of unbalanced design with unequal group sample size.

We consider two scenarios for the disturbance process $\{\boldsymbol{e}_t\} := (\boldsymbol{e}_{1t}^\top, \dots, \boldsymbol{e}_{at}^\top)^\top$: independent groups (Scenario 1) and correlated groups (Scenario 2). For both scenarios, $\{\boldsymbol{e}_t\}$ follows a multivariate moving average model. In specific, we suppose that

$$\boldsymbol{e}_t = \boldsymbol{\epsilon}_t + \boldsymbol{\Psi}\boldsymbol{\epsilon}_{t-1}, \tag{7.3}$$

where $\boldsymbol{\Psi} = (\Psi_{ij})$ is the coefficient matrix. In Scenario 1, $\boldsymbol{\Psi} = 0.7\boldsymbol{I}_a$. In Scenario 2, $\Psi_{3k-2,3k-2} = -0.5$, $\Psi_{3k,3k} = 0.4$, $\Psi_{3k,3k-2} = 0.4$, and $\Psi_{3k,3k-1} = 0.2$ for positive integer $k \leq a/3$, and $\Psi_{ij} = 0$ otherwise. The disturbance process $\boldsymbol{\epsilon}_t$ is an i.i.d. sequence, and follows the centered multivariate normal distribution with variance Σ or centered multivariate t distribution with scale Σ and 5 degrees of freedom.

For both processes of ϵ_t, in Scenario 1, Σ is an identity matrix. In Scenario 2, the diagonal element of Σ is equal to 1, and $\Sigma_{j,j+1} = \Sigma_{j+1,j} = 0.5$ for $1 \le j \le a-1$.

The following three situations for $\boldsymbol{\alpha}_i$ are considered: (A) $\boldsymbol{\alpha} = (\alpha_1, \ldots, \alpha_a)^\top = \mathbf{0}_a$ which corresponds to the null hypothesis for both fixed effect and random effect models, i.e., H_{41} and H_{42}; (B) $\boldsymbol{\alpha} = (\alpha_1, \ldots, \alpha_a)^\top$, where $\alpha_{3k-2} = 0.1, \alpha_{3k-1} = -0.1, \alpha_{3k} = 0.2$ for $k \le a/3$. This corresponds to the alternative hypothesis case for fixed effect model K_{41}; and (C) $\boldsymbol{\alpha}$ follows the centered normal distribution with variance Σ^{α}, where Σ^{α} is a block diagonal matrix whose off-diagonal blocks are all 3-by-3 zero matrix and main-diagonal blocks are all the same 3-by-3 matrix Σ^b:

$$\Sigma^b = \begin{pmatrix} 2 & 1 & 0 \\ 1 & 4 & 0.5 \\ 0 & 0.5 & 1 \end{pmatrix} \Big/ 1000.$$

The situation (C) corresponds to the alternative K_{42}.

7.2.2 Test Results

We report the rejection probabilities of the test statistic $T_{\mathrm{GALT},n}$ defined in (4.7) and the classical test $T_{\mathrm{ts},n}$ defined in (4.6) over 1000 simulations.

Figure 7.4 shows the empirical size of the tests under the null hypothesis. Both tests work relatively well for $a = 3$ and Scenario 1 (the top-left plot) for all processes. The test based on $T_{\mathrm{GALT},n}$ performs better than $T_{\mathrm{ts},n}$ for $a = 6, 9$ under Scenarios 1 and 2 if the sample size of each group is larger or equal to 2000. Under Scenario 2 with correlated groups, the test based on $T_{\mathrm{GALT},n}$ outperforms the classical test based on $T_{\mathrm{ts},n}$ with smaller size distortion for all designs; this is not surprising as the correlated groups are dealt with. On the other hand, when the sample size of each group is small, e.g., (BI) $n_1 = \cdots = n_a = 300$, the test based on $T_{\mathrm{GALT},n}$ performs worse than the classical test, and exhibits small size distortion as shown from the second to the last plot, although the performance considerably improves by increasing the sample size. The difference in performance between the two processes for ϵ_t, i.e., normal and student t, is mixed, and becomes negligible in the independent case when the sample size is large or in the correlated case.

Figure 7.5 shows the empirical power of the tests under the alternative hypothesis of fixed effects. The upper panels for $a = 3, 6, 9$ and Scenario 1 display that the empirical power of both tests are nearly equal in each model; under Scenario 2, both tests perform well with power of $T_{\mathrm{ts},n}$ being one when the sample size is larger than or equal to each other, and the power of the test based on $T_{\mathrm{GALT},n}$ is always around 1, regardless of the number of groups a, sample size, and whether the group is balanced or not. For Scenario 1, the power of $T_{\mathrm{GALT},n}$ exhibits better or similar performance with $T_{\mathrm{ts},n}$, and approaches one when the sample size is greater than 1000.

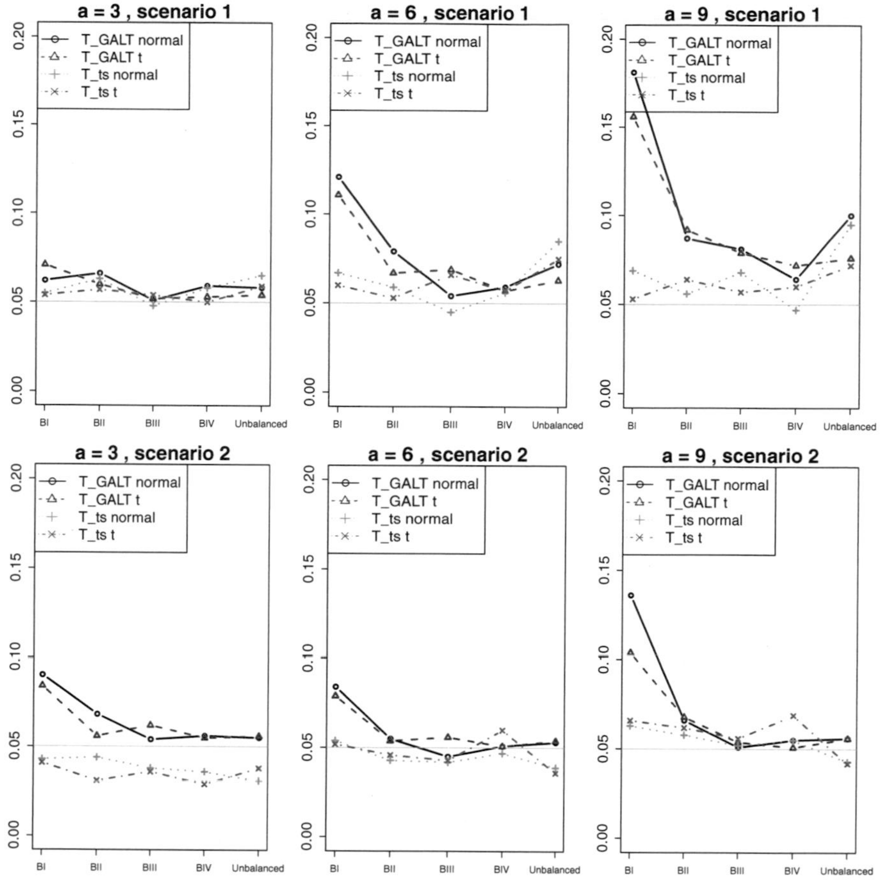

Fig. 7.4 Empirical size of tests for the existence of fixed and random effects based on $T_{\mathrm{GALT},n}$ and $T_{\mathrm{ts},n}$. The three columns correspond to $a = 3$, 6, and 9, respectively. The upper and lower panels correspond to Scenarios 1 (independent groups) and 2 (correlated groups), respectively. The tick marks of the x-label (BI), (BII), (BIII), (BIV), and (Unbalanced) correspond to the five sample size cases, respectively

Figure 7.6 shows the empirical power of the tests under the alternative hypothesis of random effects. It shows that the performance improves by increasing the sample size. Again the upper panels for $a = 3, 6, 9$ and Scenario 1 show that empirical power of both tests are similar for each model, while as expected, the lower panels highlight that the test based on $T_{\mathrm{GALT},n}$ outperforms the classical test based on $T_{\mathrm{ts},n}$ under correlated groups (Scenario 2). In most cases, the size and the power for the unbalanced design (Unbalanced) are close to the results for the balanced designs (BII) and (BIII), i.e., $n_i = 1000$ and 2000 for all groups, respectively. For almost all cases, the test based on $T_{\mathrm{GALT},n}$ is better than the classical test based on $T_{\mathrm{ts},n}$

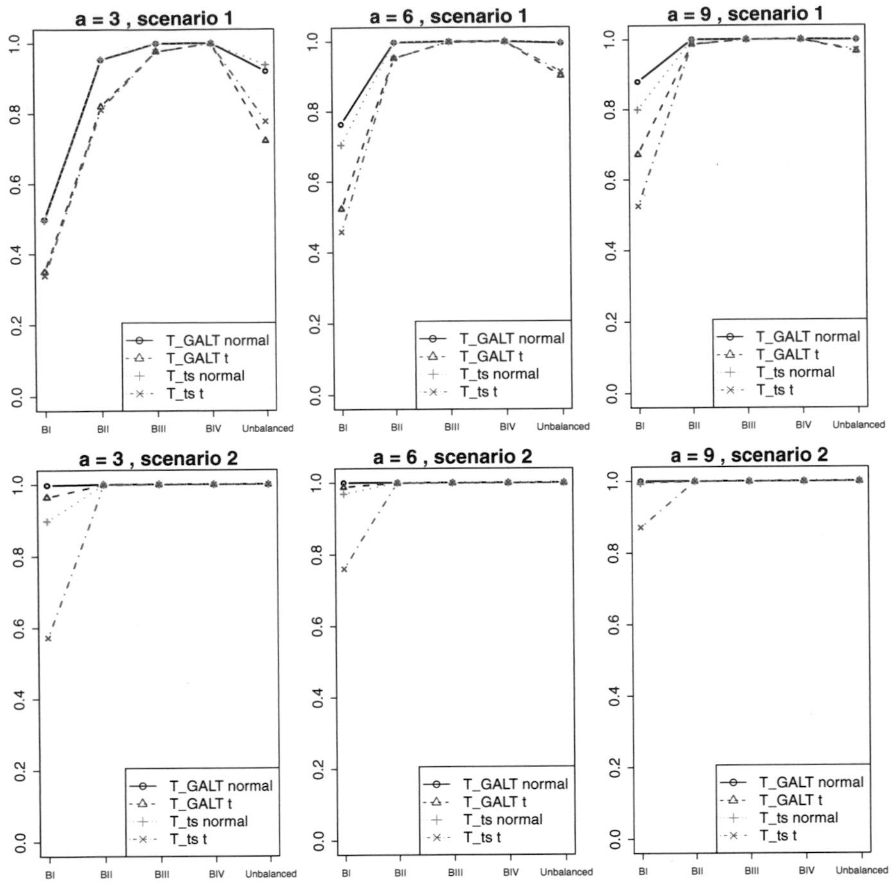

Fig. 7.5 Empirical power of tests for the existence of fixed effects based on $T_{\mathrm{GALT},n}$ and $T_{\mathrm{ts},n}$. The three columns correspond to $a = 3$, 6, and 9, respectively. The upper and lower panels correspond to Scenarios 1 (independent groups) and 2 (correlated groups), respectively. The tick marks of the x-label (BI), (BII), (BIII), (BIV), and (Unbalanced) correspond to the five sample size cases, respectively

with larger empirical power. Under the case with correlated groups (Scenario 2), the empirical power of both tests approach one when the sample size is larger than 2000 and $a = 6, 9$, as shown in the last two plots of the bottom row. The process for $\boldsymbol{\epsilon}_t$ makes a significant difference; the power under the normal distribution is better than that under the student t for all designs. Overall, the test based on $T_{\mathrm{GALT},n}$ works well in detecting the existence of fixed or random effects. In summary, the test based on $T_{\mathrm{GALT},n}$ outperforms the classical test when the groups are correlated and the sample size of each group is sufficiently large.

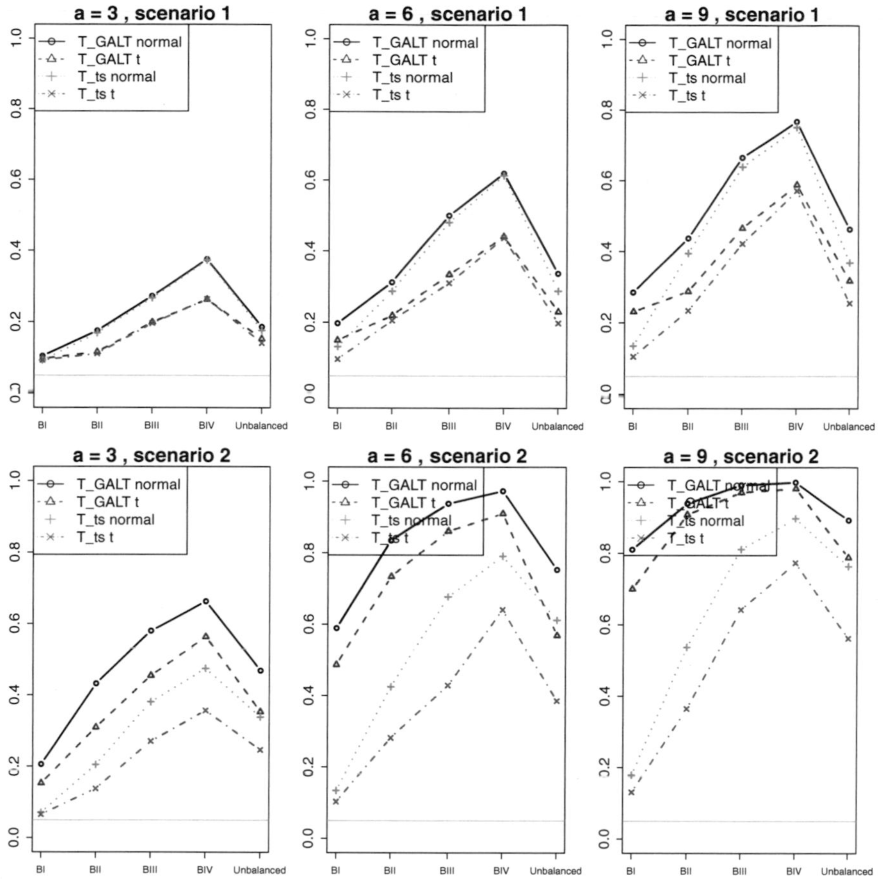

Fig. 7.6 Empirical power of tests for the existence of random effects based on $T_{\mathrm{GALT},n}$ and $T_{\mathrm{ts},n}$. The columns correspond to $a = 3$, 6, and 9, respectively. The upper and lower panels correspond to Scenarios 1 (independent groups) and 2 (correlated groups), respectively. The tick marks of the x-label (BI), (BII), (BIII), (BIV), and (Unbalanced) correspond to the five sample size cases, respectively

7.3 Two-Way Effect Model with Correlated Groups

In this part, we study the finite sample performance of the tests for random models with time-dependent errors and correlated groups. As an illustration, we investigate the performance of the test for the existence of random effects and random interactions based on the two-way random effect models in Chapter 5.

In specific, two experiments are carried out. The first deals with random effects in two-way models without interaction. The second concerns interaction effects in two-way models with interaction. The empirical size of the tests under the null hypothesis and the power under the alternative hypothesis with different types of dependent errors are investigated.

7.3.1 Test for the Existence Random Effects

In the first experiment, we focus on the tests for random effects without interaction term $\boldsymbol{\gamma}_{ij}$, that is, the data are generated from model (5.1). The parameters are $(a, b, p) = (2, 2, 1)$, the sample size is $n \in \{100, 250, 500, 1000, 2000, 3000\}$, the number of replications is $R = 1000$, the nominal level is $\tau = 0.05$, and the random effects $(\alpha_1, \alpha_2)^\top$ follow the centered bivariate normal distribution with variance ${}^{\alpha}\boldsymbol{\Sigma} = \sigma_{\alpha}^2 I_2$ and $\sigma_{\alpha} \in \{0, 0.1, 0.3, 0.5\}$ corresponding to the null and the alternative, respectively. For the innovation time series $\boldsymbol{e}_t := (e_{11t}, e_{21t}, e_{12t}, , e_{22t})^\top$, we consider vector AR(1) or MA(1) models, that is, $\boldsymbol{e}_t := \Psi \boldsymbol{e}_{t-1} + \boldsymbol{\epsilon}_t$ or $\boldsymbol{e}_t := \boldsymbol{\epsilon}_t + \Psi \boldsymbol{\epsilon}_{t-1}$, respectively. Here, $\boldsymbol{\epsilon}_t := (\epsilon_{11t}, \epsilon_{21t}, \epsilon_{12t}, \epsilon_{22t})^\top$ are i.i.d. white noise. We consider two distributions for the noise $\boldsymbol{\epsilon}_t$, the standard normal distribution and the centered t-distribution with 5 degrees of freedom and unit variance. For both cases, the coefficient matrix Ψ is defined as

$$\Psi := \begin{pmatrix} -0.4 & 0 & 0 & c_1 \\ 0 & 0.7 & 0 & 0 \\ c_2 & 0 & 0.5 & 0 \\ 0 & 0.3 & 0 & 0.3 \end{pmatrix},$$

for $(c_1, c_2) \in \{(0, 0), (-0.5, 0.2), (-0.5, 0), (-0.5, -0.5), (0.2, -0.5)\}$, respectively. The element c_1 in Ψ refers to the inter-group correlation between the coordinate (1,1)-cell and the (2,2)-cell, and c_2 indicates within-group correlation with respect to the factor A between the (1,1)-cell and the (1,2)-cell. The empirical size (under null) and the power (under alternative) are computed by

$$\frac{1}{R} \sum_{i=1}^{R} I \left\{ T^{(i)}_{\alpha,\mathrm{GSXT},n} \geq \chi^2_{\hat{r}_{n,\alpha}}[1 - \tau] \right\}, \tag{7.4}$$

where $T^{(i)}_{\alpha,\mathrm{GSXT},n}$ denotes the value of the statistic (5.5) in the i-th replication and $I\{\omega\}$ is an indicator function which takes value one when ω holds and zero otherwise.

Figures 7.7 and 7.8 display the empirical size and the empirical power of the test for the existence of random effects with different setups, and the disturbances follow the normal and the $t-$distribution, respectively. We can see that when the sample size n increases, the performance of the test improves in all the cases. From the top panel, which refers to the case under null hypothesis without random effect ($\sigma_{\alpha} = 0$), the simulated level of the test for the model with the AR(1)-type innovation approaches the nominal level 0.05 when $n = 2000$, while the empirical size of the test for the model with MA(1) error reaches the nominal level faster when n is as large as 500 and the performance is slightly more stable than with the AR(1) case. Under the alternative, the results show that the test under both AR(1) and MA(1) errors exhibits increasing power with n, and the test shows a larger power with MA(1)-type innovation than with the AR(1)-type innovation at small sample size and greater σ_{α}.

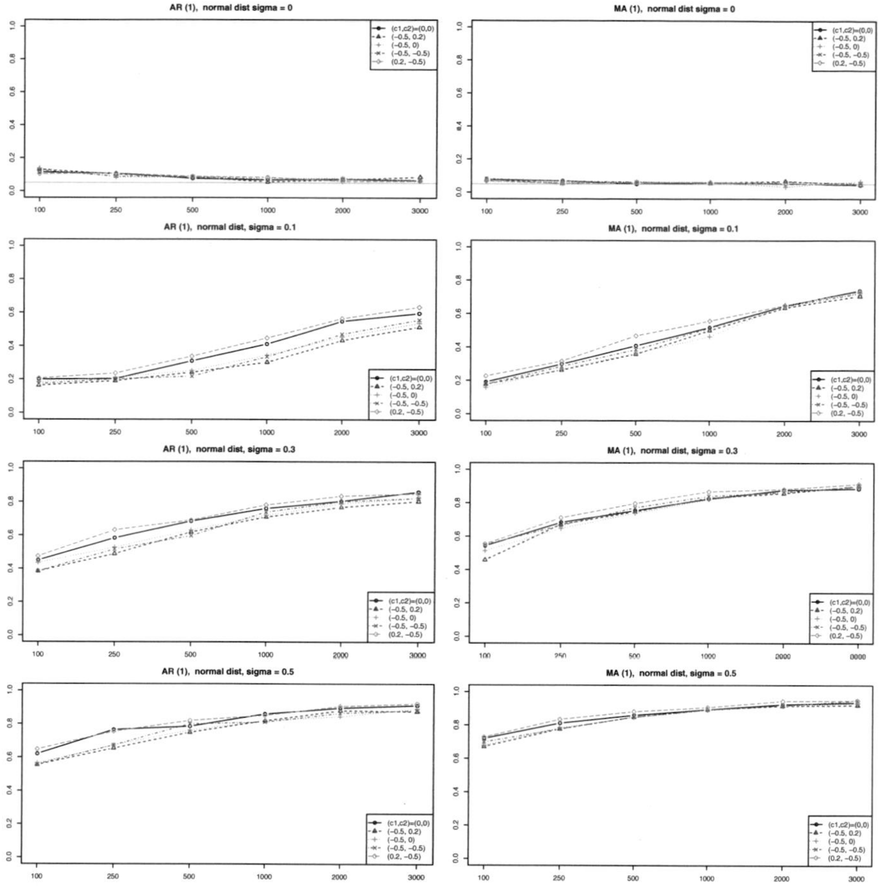

Fig. 7.7 Empirical size (the top row) and power (the second to bottom rows) for the test of random effects. The x-axis is sample size and y-axis is the average rejection probabilities over 1000 replications in each plot. The first panel refers to the null ($\sigma_\alpha = 0$), and the second to bottom panel corresponds to the alternative ($\sigma_\alpha = 0.1, 0.3, 0.5$). The left and right columns correspond to the AR(1)- and MA(1)-type innovations with disturbances following the normal distribution, respectively

As concerns the distributions of the disturbances $\boldsymbol{e}_t$, our test shows that there is no significant difference between the normal and the student t-distribution under the null, whereas under the alternative, the power is slightly higher when $\boldsymbol{e}_t$ follows the normal distribution rather than the t distribution.

As for σ_α, it can be seen from the second, third, and fourth columns that the power of our test becomes larger when σ_α increases. For example, when $\sigma_\alpha = 0.1$, the power of the test is close to 0.6 when $n = 3000$ for the case of AR(1)-type innovation and normal disturbances, while the power increases to 0.9 when σ_α becomes 0.5. This is not surprising, as larger σ_α indicates more significant random effects, which augment the power of the test.

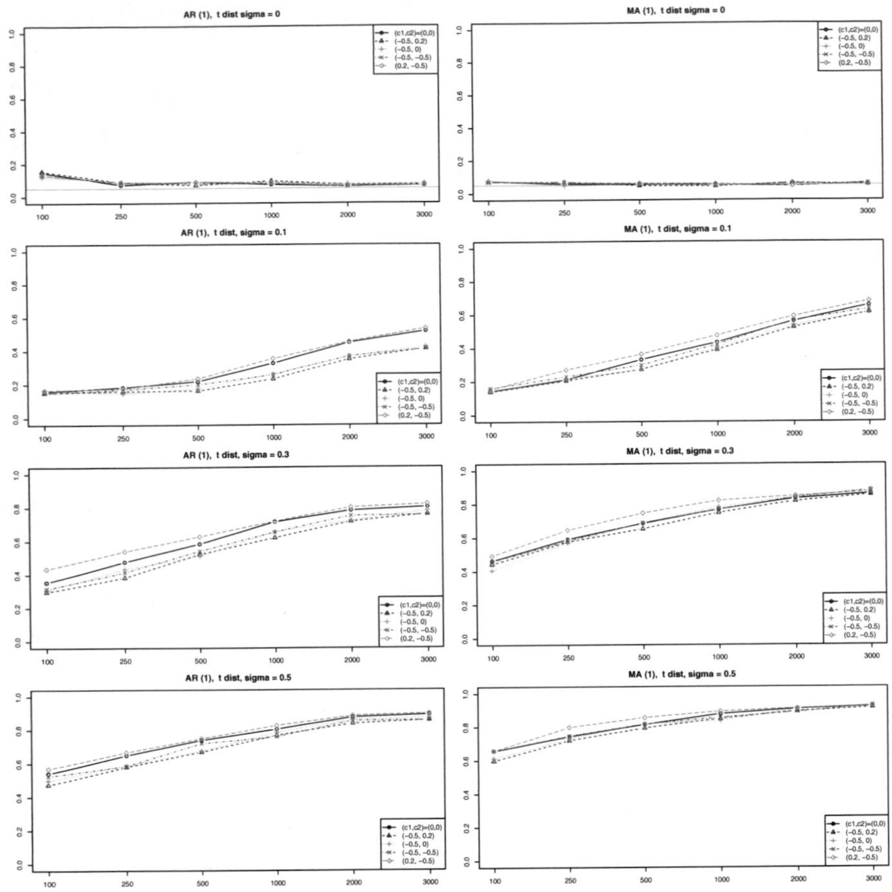

Fig. 7.8 Empirical size (the top row) and power (the second to bottom rows) for the test of random effects. The x-axis is sample size and y-axis is the average rejection probabilities over 1000 replications in each plot. The first panel refers to the null ($\sigma_{\alpha} = 0$), and the second to bottom panel corresponds to the alternative ($\sigma_{\alpha} = 0.1, 0.3, 0.5$). The left and right columns correspond to the AR(1)- and MA(1)-type innovations with disturbances following the t-distribution, respectively

7.3.2 *Test of Interaction Effects*

In this subsection, we include nonzero interaction γ_{ij} in the two-way effect model and generate data from model (5.6). We consider the same settings as in Section 7.3.1, except that the interactions term $(\gamma_{11}, \gamma_{21}, \gamma_{12}, \gamma_{22})^{\top}$ follows the centered multivariate normal distribution with variance ${}^{\gamma}\boldsymbol{\Sigma} = \sigma_{\gamma}^2 I_4$ for $\sigma_{\gamma} \in \{0, 0.1, 0.3, 0.5\}$. Here, the case $\sigma_{\gamma} = 0$ and nonzero σ_{γ} correspond to the null and the alternative hypothesis, respectively. Similarly, we define the rejection probability, i.e., the empirical size (under null) and the power (under alternative) by

$$\frac{1}{R}\sum_{i=1}^{R} I\left\{T^{(i)}_{\gamma,\mathrm{GSXT},n} \geq \chi^2_{\hat{r}_{n,\gamma}}[1-\tau]\right\}, \tag{7.5}$$

where $T^{(i)}_{\gamma,\mathrm{GSXT},n}$ denotes the value of the test statistic defined in (5.8) in the i-th replication.

Figures 7.9 and 7.10 show the empirical size and power of the tests for ${}^{\gamma}\boldsymbol{\Sigma}$ with the disturbances following the normal and the t−distribution, respectively. The results

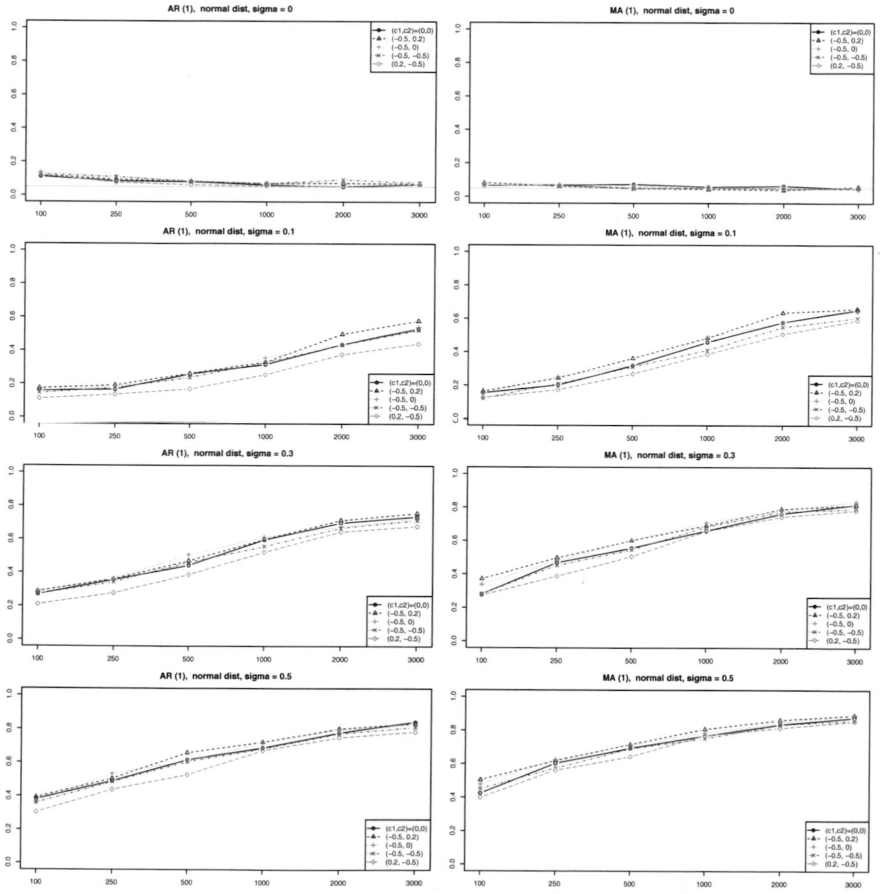

Fig. 7.9 Empirical size (the top row) and power (the second to bottom rows) for the test of interaction effects. The x-axis is sample size and y-axis is the average rejection probabilities over 1000 replications in each plot. The first panel corresponds to the null ($\sigma_\gamma = 0$), and the second to bottom panel refers to the alternative ($\sigma_\gamma = 0.1, 0.3, 0.5$). The left and right columns correspond to the AR(1)- and MA(1)-type innovations with disturbances following the normal distribution, respectively

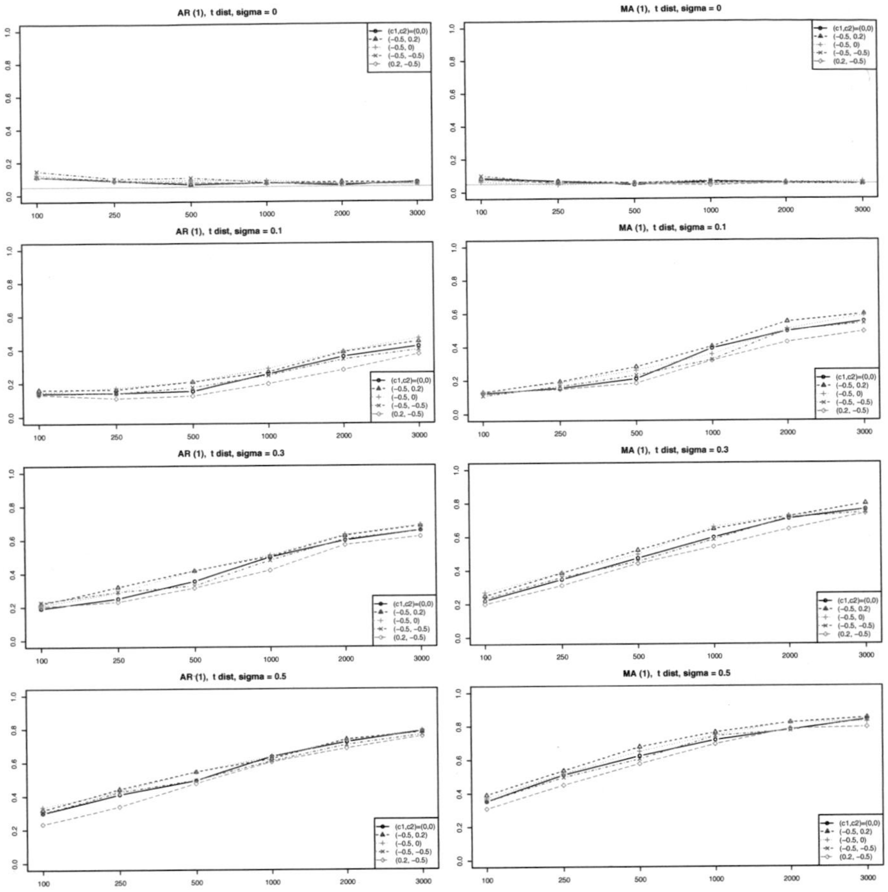

Fig. 7.10 Empirical size (the top row) and power (the second to bottom rows) for the test of interaction effects. The x-axis is sample size and y-axis is the average rejection probabilities over 1000 replications in each plot. The first panel corresponds to the null ($\sigma_\gamma = 0$), and the second to bottom panel refers to the alternative ($\sigma_\gamma = 0.1, 0.3, 0.5$). The left and right columns correspond to the AR(1)- and MA(1)-type innovations with disturbances following the t-distribution, respectively

are similar to the previous experiment, that is, the test performance improves with larger sample size, and the test for the model with MA(1) error reaches the nominal level faster and generally exhibits higher power than in the case of AR(1)-type innovation. No significant difference is observed between the normal and the student t-distributed noise under the null. The power of the test increases with σ_α which refers to higher interaction effects in the two-way models.

Appendix: DCC-GARCH Parameters

DCC-GARCH $p = 2$

$$\sigma_{it}^{(1)} = 0.001 + 0.05\left[\epsilon_{it}^{(1)}\right]^2 + 0.90\sigma_{i(t-1)}^{(1)}$$
$$\sigma_{it}^{(2)} = 0.001 + 0.10\left[\epsilon_{it}^{(1)}\right]^2 + 0.85\sigma_{i(t-1)}^{(2)}$$
$$\tilde{\boldsymbol{Q}} = \begin{bmatrix} 1.0 & 0.5 \\ 0.5 & 1.0 \end{bmatrix}.$$

DCC-GARCH $p = 3$

$$\sigma_{it}^{(1)} = 0.001 + 0.05\left[\epsilon_{it}^{(1)}\right]^2 + 0.92\sigma_{i(t-1)}^{(1)}$$
$$\sigma_{it}^{(2)} = 0.001 + 0.10\left[\epsilon_{it}^{(1)}\right]^2 + 0.85\sigma_{i(t-1)}^{(2)}$$
$$\sigma_{it}^{(3)} = 0.001 + 0.07\left[\epsilon_{it}^{(1)}\right]^2 + 0.90\sigma_{i(t-1)}^{(3)}$$
$$\tilde{\boldsymbol{Q}} = \begin{bmatrix} 1.0 & 0.5 & 0.4 \\ 0.5 & 1.0 & 0.3 \\ 0.4 & 0.3 & 1.0 \end{bmatrix}.$$

DCC-GARCH $p = 4$

$$\sigma_{it}^{(1)} = 0.001 + 0.05\left[\epsilon_{it}^{(1)}\right]^2 + 0.92\sigma_{i(t-1)}^{(1)}$$
$$\sigma_{it}^{(2)} = 0.001 + 0.10\left[\epsilon_{it}^{(1)}\right]^2 + 0.85\sigma_{i(t-1)}^{(2)}$$
$$\sigma_{it}^{(3)} = 0.001 + 0.07\left[\epsilon_{it}^{(1)}\right]^2 + 0.90\sigma_{i(t-1)}^{(3)}$$
$$\sigma_{it}^{(4)} = 0.001 + 0.06\left[\epsilon_{it}^{(1)}\right]^2 + 0.88\sigma_{i(t-1)}^{(4)}$$
$$\tilde{\boldsymbol{Q}} = \begin{bmatrix} 1.00 & 0.45 & 0.40 & 0.50 \\ 0.45 & 1.00 & 0.30 & 0.73 \\ 0.40 & 0.30 & 1.00 & 0.40 \\ 0.50 & 0.73 & 0.40 & 1.00 \end{bmatrix}.$$

DCC-GARCH $p = 5$

$$\sigma_{it}^{(1)} = 0.001 + 0.05\left[\epsilon_{it}^{(1)}\right]^2 + 0.92\sigma_{i(t-1)}^{(1)}$$
$$\sigma_{it}^{(2)} = 0.001 + 0.10\left[\epsilon_{it}^{(1)}\right]^2 + 0.85\sigma_{i(t-1)}^{(2)}$$
$$\sigma_{it}^{(3)} = 0.001 + 0.07\left[\epsilon_{it}^{(1)}\right]^2 + 0.90\sigma_{i(t-1)}^{(3)}$$
$$\sigma_{it}^{(4)} = 0.001 + 0.06\left[\epsilon_{it}^{(1)}\right]^2 + 0.88\sigma_{i(t-1)}^{(4)}$$
$$\sigma_{it}^{(5)} = 0.001 + 0.08\left[\epsilon_{it}^{(1)}\right]^2 + 0.92\sigma_{i(t-1)}^{(5)}$$
$$\tilde{\boldsymbol{Q}} = \begin{bmatrix} 1.00 & 0.47 & 0.43 & 0.50 & 0.50 \\ 0.47 & 1.00 & 0.37 & 0.73 & 0.45 \\ 0.43 & 0.37 & 1.00 & 0.40 & 0.65 \\ 0.50 & 0.73 & 0.40 & 1.00 & 0.50 \\ 0.50 & 0.45 & 0.65 & 0.50 & 1.00 \end{bmatrix}.$$

Reference

Engle, R. (2002). Dynamic conditional correlation: A simple class of multivariate generalized autoregressive conditional heteroskedasticity models. *J. Bus. Econ. Stat.*, *20*, 339–350.

Chapter 8
Empirical Data Analysis

This chapter analyzes the average wind speed data observed in seven cities located in coastal and inland areas in Japan by assuming the one-way effect model with time-dependent errors and correlated groups. The purpose is to assess the existence of area effects. To this end, both the $T_{\text{GALT},n}$ statistic designed for correlated groups and the $T_{\text{ts},n}$ statistic designed for independent groups are applied. The results highlight the greater accuracy of the former test which appropriately takes into account the correlation structure among the series.

8.1 Average Wind Speed Data

In this chapter, we carry out some real data analysis to investigate the finite sample performance of the tests for one-way models presented in the previous chapters. In specific, we illustrate the performance of the test in the one-way effect model with time-dependent errors and correlated groups based on model (4.1) of Chapter 4.

We consider the average wind speed data observed in seven cities (points) in Japan: Matsumoto, Maebashi, Kumagaya, Chichibu, Tokyo, Chiba, and Yokohama. The location map of the seven points is displayed in Fig. 8.1. Three cities are located in coastal areas (Tokyo, Chiba, and Yokohama), while the other four cities are in inland areas. The data come from the Japan Meteorological Agency (JMA), and can be found at https://www.data.jma.go.jp/gmd/risk/obsdl/index.php. The JMA records weather observations for the purpose of preventing weather disasters by issuing domestic weather forecasts, warnings, and advisories. An additional purpose is to monitor the climate for the development of industry and the conservation of the global environment. The JMA collects automatic and visual observations of atmospheric pressure, wind, and other meteorological phenomena at approximately 150 observation facilities nationwide. In addition, in order to grasp wind and other meteorological phenomena in more detail, the JMA automatically observes wind speed at approximately 1,300 observation stations throughout Japan.

Y. Goto et al., *ANOVA with Dependent Errors*, JSS Research Series in Statistics,
https://doi.org/10.1007/978-981-99-4172-8_8

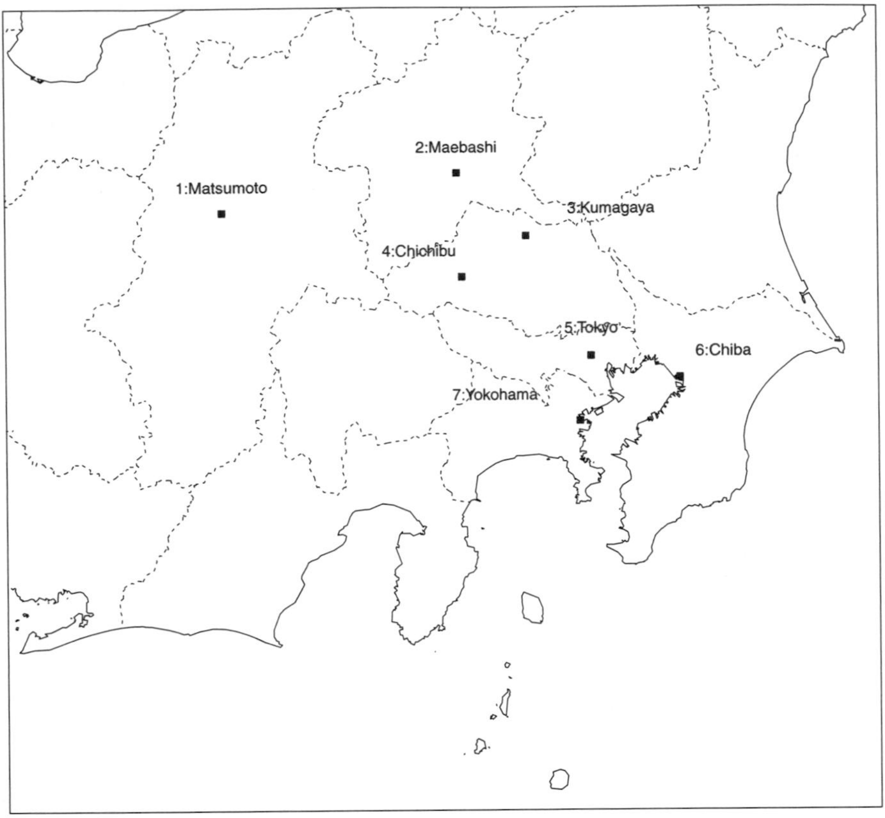

Fig. 8.1 Map of seven cities and corresponding locations

Table 8.1 The mean, minimum value, maximum value, and variance of daily average wind speed data from May 1, 2020, to April 1, 2023 (1,066 observations)

Prefectures	Matsumoto	Maebashi	Kumagaya	Chichibu	Tokyo	Chiba	Yokohama
Mean	2.50	2.30	2.42	1.62	2.70	3.63	3.46
Min	0.50	0.70	0.90	0.50	1.20	1.20	1.30
Max	7.00	6.40	7.10	4.20	6.50	11.20	8.90
Variance	1.33	0.75	0.98	0.36	0.71	2.28	1.41

We apply the test statistics to the average wind speed series daily recorded from May 1, 2020, to April 1, 2023. Hence, the dataset consists of daily average wind speed series from seven points covering 1,066 observations for each city. Table 8.1 reports the mean, minimum value, maximum value, and variance of the average wind speed time series data for each point. It can be seen that the average wind speeds in coastal prefectures (Tokyo, Chiba, and Yokohama) are higher than those in inland prefectures (Matsumoto, Maebashi, Kumagaya, and Chichibu) with a greater

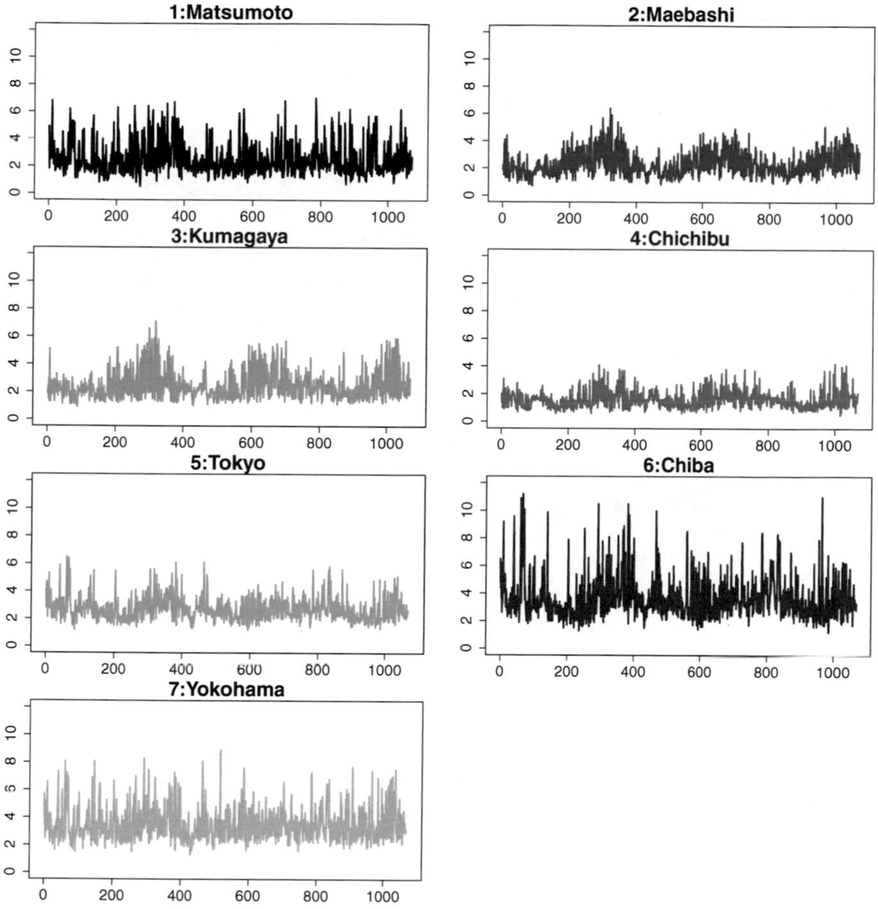

Fig. 8.2 Time series plots of the average wind speed data of seven points from May 1, 2020, to April 1, 2023

magnitude of the mean. It is also interesting to note that wind speeds in Chichibu are smaller than those in other inland prefectures.

Figure 8.2 displays the time series plots of the average wind speed data of the seven points from May 1, 2020, to April 1, 2023. All the series are relatively stationary with a slightly increasing trend since the middle of year 2022, and the patterns of these observation sequences are similar.

Figure 8.3 displays the heatmap plot of correlation of the average wind speed series among the seven points. It can be seen that the correlations of the average wind speeds for both coastal prefectures and inland areas are high except for Matsumoto. This may be due to the close distance between the observation sites, whereas Matsumoto is located relatively far from the other observation sites. The significant correlation shown in the heatmap indicates that the test based on $T_{\mathrm{GALT},n}$ designed for correlated

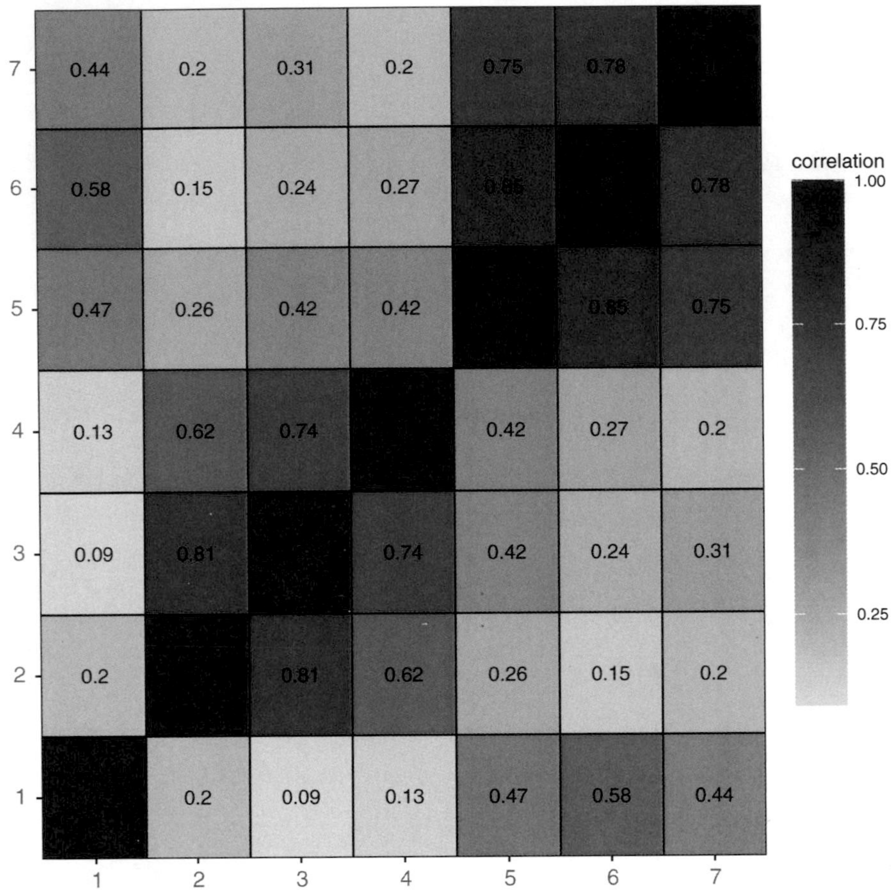

Fig. 8.3 The heatmap of sample correlations between points

groups may be more appropriate than that based on $T_{\text{ts},n}$ designed for independent groups.

We applied the test statistic $T_{\text{GALT},n}$ for the one-way effect model with correlated groups to the seven average wind speed series. Each city is located at different prefectures and is considered as a group, referring to the area effect. As a comparison, we also carried out the classical test based on $T_{\text{ts},n}$ which is designed for independent groups. The values of the statistics are 1796 and 973, and the corresponding p-values are 0 and 5.22×10^{-207} for $T_{\text{GALT},n}$ and $T_{\text{ts},n}$, respectively. This indicates that there is strong evidence to reject the null hypothesis that there is no area effect for both tests. However, $T_{\text{ts},n}$ provides a low value of 973 compared with 1796 of $T_{\text{GALT},n}$, indicating that ignoring the between-group correlation may cause misleading results. This analysis points out that it is crucial to apply tests that are suited to the features of the data to ensure reliable conclusions.

Index

Y. Goto et al., *ANOVA with Dependent Errors*, JSS Research Series in Statistics,
https://doi.org/10.1007/978-981-99-4172-8

Printed in the United States
by Baker & Taylor Publisher Services